AF400248

Supplement to
Basic Research in Cardiology, Vol. 89, Suppl. 1 (1994)
Editor:
G. Heusch, Essen

H. Just, W. Hort, A. M. Zeiher (Eds.)

Arteriosclerosis

New Insights into Pathogenetic Mechanisms and Prevention

 Steinkopff Verlag Darmstadt
Springer-Verlag New York 

The Editors:

Prof. Dr. H. Just
Medizinische Universitätsklinik
Abt. Innere Medizin III
Hugstetter Straße 55
79106 Freiburg

PD Dr. A. M. Zeiher
Medizinische Universitätsklinik
Abt. Innere Medizin III
Hugstetter Straße 55
79106 Freiburg

Prof. Dr. W. Hort
Pathologisches Institut der
Heinrich-Heine-Universität
Moorenstraße 5
40225 Düsseldorf

Die Deutsche Bibliothek – CIP-Einheitsaufnahme
[Basic research in cardiology / Supplement]
Supplement to Basic research in cardiology. – Darmstadt :
Steinkopff ; New York : Springer.
   Teilw. nur im Verl. Steinkopff, Darmstadt. – Früher Schriftenreihe
   Fortlaufende Beil. zu: Basic research in cardiology
NE: HST
Vol. 89, Suppl. 1. Arteriosclerosis. – 1994
Arteriosclerosis : new insights into pathogenetic mechanisms
and prevention / H. Just ... (ed.). – Darmstadt : Steinkopff ;
New York : Springer, 1994
   (Supplement to Basic research in cardiology ; Vol. 89, Suppl. 1)
   ISBN-13: 978-3-642-85662-4   e-ISBN-13: 978-3-642-85660-0
   DOI: 10.1007/978-3-642-85660-0
NE: Just, Hansjörg [Hrsg.]

Suppl. Basic Res. Cardiol, ISSN 0175-9418
Indexed in Current Contents.

This work is subject to copyright. All rights are reserved, whether the whole or part of the material is concerned, specifically the right of translation, reprinting, re-use of illustrations, recitation, broadcasting, reproduction on microfilms or in other ways, and storage in data banks. Duplication of parts thereof is only permitted under the provisions of the German Copyright Law of September 9, 1965, in its version of June 24, 1985, and a copyright fee must always be paid. Violations fall under the prosecution act of the German Copyright Law.

Copyright © 1994 by Dr. Dietrich Steinkopff Verlag GmbH & Co. KG, Darmstadt
Softcover reprint of the hardcover 1st edition 1994
Medical Editor: Sabine Müller – English Editor: James C. Willis – Production: Heinz J. Schäfer

The use of registered names, trademarks, etc. in this publication does not imply, even in the absence of a specific statement, that such names are exempt from the relevant protective laws and regulations and therefore free for general use.

Typesetting: Typoservice, Alsbach

Printed on acid-free paper

Introduction

A vast literature has been concerned with arteriosclerosis and yet, many aspects of pathogenesis and of the mechanism of development of the arteriosclerotic vascular lesion remain only poorly understood. In recent years, our knowledge of the earliest stages of arteriosclerosis have greatly improved. By now, we have learned to relate morphologic changes to disturbances in function. It has been of particular importance that components of the arterial wall could be analyzed in regard to dysfunction, for example, in the endothelium or the vascular smooth muscle. The interaction of the different morphological components of the vascular wall could thus be much better understood. Likewise, the interaction between the arterial wall and the flowing blood could be much better described, including the intimate relationship between platelets and the endothelium, the coagulation system and the endothelium, the granulocytes and the endothelial cell layer, as well as processes of migration of blood cells into the subendothelial space.

The recognition of functional and morphological disturbance has attained clinical significance not only because the arteriosclerotic diseases have quantitatively reached the dimensions of an epidemic, that is, of a magnitude never been witnessed. It is also because of the development of new drugs that interfere with the atherogenic process and thereby prevent the development of the disease or halt its progression. It is also becoming increasingly possible to inhibit the occurrence of complications in existing arteriosclerotic lesions in manifest disease, i.e., the occurrence of thrombosis, plaque rupture, and vascular occlusion. The advent of interventional catheter-therapy has brought about yet another important aspect, i.e., the occurrence of restenosis after successful dilatation or recanalization of an obstructed vessel.

All these considerations are of particular relevance and of special difficulty in the case of the coronary circulation. Here, the arteriosclerotic process develops in relatively small vessels supplying an organ which vitally depends upon continuous supply and is directly linked to the most vital functions of the organism. Indeed, the progress of pharmacotherapy and of surgical and interventional technologies has been rather dramatic in this field, and is important for future progress.

Progress can be expected in the field of clinical diagnosis of coronary disease: It has become evident that functional disturbances of the endothelium have the greatest influence upon myocardial blood supply in all stages of the disease, especially in the early stages when angiography is not yet able to detect the atheromatous process. New diagnostic techniques have been developed to detect and to quantify dysfunction. Among the pharmacotherapeutic advances are notably the calcium antagonists, as well as the inhibitors of the angiotensin converting enzymes. Our hope to halt progression and to prevent complications has been connected with these two important groups of drugs. Lipid-lowering therapy has been and is being tested as the major approach for primary and secondary prevention.

It was therefore deemed necessary to review the current state of the art of our understanding of arteriosclerosis and to outline current developments and frontiers.

To this end, a group of scientists assembled in July 1992, at a Gargellen Conference that united basic scientists and clinicians.

The contributions assembled in this supplement to *Basic Research in Cardiology* represent the proceedings of a highly stimulating symposium. The editors owe thanks to the authors for competently and clearly describing their field, thereby giving us an account of the current knowledge about morphology and pathophysiology of the arteriosclerotic disease, with special reference to the coronary circulation. The outstanding authors and the themes of their work include:

W. Hort, Düsseldorf, FRG, gives a historical review of Rokitansky, Virchow, and Langhans and describes the pathogenesis of arteriosclerosis. He then discusses unresolved problems, especially regarding localization and distribution of the disease within the arterial system. He gives an account of the morphology of the endothelium and its relation to other cellular constituents of the arterial wall.

H. Stary, New Orleans, USA, describes structure and components of atherosclerotic lesions at different ages. He gives a classification of eight lesion types characterized by cell composition, matrix, tissue architecture, and other specific features. Types I – IV are the lesions to be found in the first 4 decades of life. Type III is characteristic for adolescence and young adults. Type IV lesions are apt to produce symptoms; Types VI – VIII represent complicated stages of arterial disease with fibroatheroma, thrombotic deposits and ulcerations, always associated with the potential of severe complications.

M. Davies and coworkers from London, UK, describe the cellular and lipid components of unstable human aortic plaques. The understanding of the morphology and of the dynamics of the arteriosclerotic plaques is of greatest importance, since the majority of complications in coronary arteriosclerosis arises from plaque rupture.

G. Hansson from Gotenburg, Sweden, observes cellular components of inflammation and of immune response like T lymphocytes and cytokine secretion. T-cell clones, obtained from atherosclerotic lesions have given us new insides into the activation of immunological control mechanism in plaque formation. Vascular effects of cytokines reduced by macrophages and lymphocytes show the importance of these mechanisms in the development of the disease, as well as in the occurrence of complications.

H. E. Schäfer, Freiburg, FRG, discusses the still unresolved riddle of the role of the smooth muscle cells in human arteriosclerosis. Many an observation seems to point to the fact that atrophy of the smooth muscle cell structures prevails in arteriosclerosis. A local proliferation of the myointimal or Langhans-cells, which are closely related to the smooth muscle cells, presents an interesting but, functionally, only an incompletely understood aspect of arteriosclerosis.

A. Newby and coworkers from Cardiff, UK, discuss metalloproteinases degrading structures of the extracellular matrix. They present evidence that here an important aspect of the pathogenesis of atherosclerosis may be found. They show direct evidence that metalloproteinases are involved in the proliferation and outgrowth of vascular smooth muscle cells. Degradation of basement membrane components seems to be a prerequisite for the proliferation and outgrowth of these cells.

After discussing newer aspects of the morphology and of the cellular components of the arterial walls, J. Holtz and R. Goetz from Halle, FRG, present their view on the vascular renin-angiotensin system, function of the endothelium and atherosclerosis. They describe the central and crucial role of the endothelium as major target and seat of the angiotensin system. They present evidence that this neurohumoral system in its activated state has atherogenic effects, whereby its inhibition may produce the opposite effect.

Endothelial dysfunction is the topic of the following contributions. D. G. Harrison, Atlanta, USA, gives an overview of the function of the endothelium and its disturbances in hypercholesterolemia and arterial disease. The dynamic adaptation of vascular size to blood flow rests upon the astounding nitric oxide-dependent mechanism. This mechanism that is closely related to the bradykinin system plays a crucial role. Its multiple facets relating to cholesterol-induced damage as well as to the coagulation mechanisms in the flowing blood present the basis for the recent thrust in vascular research.

G. V. R. Born from London, UK, describes mechanisms of uptake of atherogenic plasma proteins by the arteries. The mechanism of development of atherosclerosis in hypercholesterolemia is now better understood. We know that trans-endothelial transport of lipids and their reaction with cellular components in the subendothelial space presents yet another important component besides the direct injury to the endothelial cell layer itself.

P. D. Henry of Houston, Texas, describes the relationship between endothelial dysfunction under the influence of hyperlipidemic states and processes promoting vascular growth. Endothelial replication, necessary for vascular growth, is markedly impaired in the presence of hypercholesterolemia. Ample evidence has been presented that endothelial function by means of nitric oxide production in response to blood flow presents a major vascular growth factor as well.

A. Zeiher and coworkers from Freiburg, FRG, discuss the functional disturbance of the endothelium and the entire vascular wall in the coronary circulation. Here, a new aspect of the dynamic regulation of coronary blood flow and its impairment in arteriosclerosis is described. These changes blend into a newer, more complete hypothesis of the pathogenesis of coronary arteriosclerosis. The authors describe the need for and the modalities of measurement of these functional disturbances in vivo.

K. M. Schmid and coworkers from Tübingen, FRG, present yet another technique for the description of arterial disease, i.e., the assessment of the coronary artery wall by intravascular, high-frequency ultrasound. This technique promises a better understanding of this disease in the future and may provide an aid in the planning of intervention within the coronary arterial system, as well as in larger arteries.

In the final section therapeutic approaches are discussed:

W. G. Nayler, Melbourne, Australia, presents current approaches to the control of atherosclerosis, especially using calcium antagonists and antilipidemic agents.

The particularly fascinating role of calcium in the pathogenesis of arteriosclerosis is described by G. Fleckenstein-Grün and coworkers from Freiburg, FRG, from the Study-Group for Calcium Antagonism, formerly headed by Albrecht Fleckenstein. This is a most complete overview of our current knowledge of calcium and its role in arteriosclerosis.

Then, H. Just and M. Frey from Freiburg analyze if the expectations for the application of calcium antagonists in the treatment of arteriosclerosis have been fulfilled. The conclusion is that, in certain cases, preventive effects can be seen and that in others complications of the arteriosclerotic process can be halted. On the whole, however, the possibilities to influence the dynamic progression of arteriosclerosis have been rather small.

G. Schmitz and K. J. Lackner from Regensburg, FRG, summarize our current knowledge regarding the results of lipid-lowering therapy in the prevention and regression of arteriosclerosis. Here, a significant therapeutic effect can be achieved.

The basis for these considerations are animal studies and epidemiological surveys. Only recently have interventional studies in man become available. Different strategies for intervention have been developed and can be applied to the different manifestations of the disease with good expectations for therapeutic effects in certain subgroups of patients. On the whole, our capability to halt the disease is still limited, but for subsets of patients a very remarkable reduction in complication and progression of the disease can be achieved. We therefore hope that ongoing research will lead us to more effective avenues of therapeutic control of arteriosclerosis.

The Gargellen Conference was generously supported by Bayer AG, Leverkusen, FRG. In particular, I thank Georg Bertschik for continuous support and understanding of the relevance of basic science and for the understanding of clinical phenomena. The support of Bayer has never been based on only its own products, but rather from interest in the advancement of science. Without this generous support, I would not have been able to assemble outstanding scientists from around the world.

We have again had the pleasure to work with the publisher Dr. Dietrich Steinkopff Verlag, a division of Springer-Verlag Group. Especially, we are grateful to Sabine Müller, who has given us her skillful, careful, and charming help.

The conference was organized by the Society for Cooperation in Medical Sciences, a non-profit organization aimed at linking basic science and clinical medicine in the interest of advancement of science and clinical medicine. We hope that the reader will enjoy this state-of-the-art assessment of currently important disease. The hopeless situation of yesteryear for the clinician confronted with inevitably progressing arteriosclerosis is yielding to successful modalities of primary and secondary prevention and to the treatment of typical and frequently life-threatening complications.

H. Just, Freiburg   W. Hort, Düsseldorf   A. Zeiher, Freiburg

Contents

Arteriosclerosis: Its morphology in the past and today

W. Hort

Pathologisches Institut der Heinrich-Heine-Universität Düsseldorf, FRG

Summary: In a brief historical review the contributions of Rokitansky, Virchow, and Langhans concerning the pathogenesis of arteriosclerosis and the histogenetic puzzle of intimal cells classification are described. Then, some unresolved problems are discussed, especially the localization and distribution of arteriosclerotic plaques, the shape of endothelial cells and the orientation of their nuclei in correlation to local hemodynamic stress under normal and pathologic conditions. Some differences between experimental arteriosclerosis and arteriosclerosis in humans are illustrated by examples.

The key role of the endothelium in the development of arteriosclerosis is well founded. According to recent investigations some cells on the surface of human arteriosclerotic plaques appear to be of non-endothelial origin.

Arteriosclerosis seems to be a systemic disorder with multiorgan involvement. Individual cases, however, show significant differences in the distribution and extent of lesions.

Today, arteriosclerotic research is focused on arteries being most important in clinical investigations. Nevertheless, there are also other arteries with severe arteriosclerotic lesions; for example, the degree of arteriosclerosis in periprostatic arteries is more pronounced than in coronary artery branches of the same size.

Finally, the importance of primary prevention of arteriosclerosis is emphasized.

Key words: Historical review – endothelium – intimal cells – development and localization of arteriosclerotic plaques

Introduction

Almost one and a half centuries ago – in 1845 – Friedrich Wilhelm IV, King of Prussia, visited his old friend Prof. Argelander, who was a famous astronomer at Bonn. The king asked the scientist, "What is new in the sky?" Prof. Argelander replied, "Does Your Majesty already know all the old things?"

You have come to Gargellen to discuss some of the latest findings about arteriosclerosis. But according to the spirit of the reply of Prof. Argelander, let us first review some aspects of the past. Afterwards, we will point to some contemporary, unresolved morphologic problems concerning arteriosclerosis.

At first, let us try to reconstruct the morphologic knowledge of arteriosclerosis in 1845, the year of the mentioned meeting of the king and the scientist at Bonn.

Twelve years earlier, in 1833, Lobstein (21) had created the term "arteriosclerosis" and he explained this name by the following footnote: "Nom compose d'artère et de sclerosis, épaissement avec induration" (= composed name from artery and sclerosis, condensation and induration). Lobstein was a pathologist at Strasbourg and there he held the first chair of pathology established in Europe. In his

"Traité d'anatomie pathologique" he introduced a new arrangement of the matter, following anatomical aspects.

In 1845, Rokitansky (Fig. 1) was 41 years old. By that time he was already the leading pathologist in Europe and the head of the famous new medical school of Vienna. A large part of his Manual of Pathologic Anatomy arranged in Lobstein's manner had just been published. Rokitansky was outstanding in his ability to describe and in his vividness. In his chapter on arteries (26), he described a deposition of an inner vascular membrane as being by far the most frequent disease of arteries, leading to the formation of aneurysms and of many spontaneous obliterations. He deduced that the deposit derived from the arterial blood. This statement

Fig. 1. Carl von Rokitansky (1804–1878)

2

can be understood in connection with the attempt of Rokitansky to revive the idea of dyscrasia, which had been popular already in antiquity.

Today, Rokitansky is looked upon as the father of the thrombogenic or incrustation theory of arteriosclerosis. But in his lifetime, his theory of dyscrasia was vehemently attacked by Virchow.

In 1845, the star of Rudolf Virchow (Fig. 2) began to rise. He was only 24 years old and was the assistant of Froriep, prosector at the Charité at Berlin. In the following year, Virchow succeeded Froriep as prosector and he rejected the views of Rokitansky concerning the patogenesis of arteriosclerosis (40). The essential instrument of Virchow was the microscope, whereas the strength of Rokitansky was macroscopic observation. Virchow regarded the cell as the center of illness, and in 1858, he published his epoch-making "Cellularpathologie" (42).

In this work, Virchow (42) described that the intimal thickening in arteriosclerosis was located in the subendothelial layer and, therefore, he concluded that it could not be derived from surface deposits. These findings were convincing, and in 1855, in the third edition of his handbook, Rokitansky (27) largely revoked his theory of deposi-

Fig. 2. The young Rudolf Virchow (From Virchows Arch Abt. A [1971], Heft 4, Zum 150. Geburtstag von Rudolf Virchow)

tion on the surface of arteries. Therefore, a possible relationship between thrombosis and plaque formation was neglected for a long time.

In the meantime, the reputation of Rokitansky decreased. Now Virchow was the leading pathologist, and in Central Europe, the leading medical man, too.

What was known about the structure of the arterial wall in the time of the controversy between Virchow and Rokitansky? The inner surface of blood vessels was believed to be covered by an epithelium. The endothelium was unknown. In 1840, Henle (15) described smooth muscle cells in the media.

Remember that, in the middle of the last century, methods for microscopic investigations were still rather limited. The microtome was not available until 1954 (see 8). Before this time, thin particles were produced by pulling, pressing or cutting with a razor, and macroscopic preparation was of great importance. In preparing the intima, frequently superficial layers can be easily pulled of and can be taken for deposits. Therefore, Rokitansky's idea of depositions on the surface of the intima was quite plausible.

The intimal cells have a celebrated past history. Important contributions derived from Rokitansky (27) and Virchow (42). They described and illustrated cells of the intima in arteriosclerotic lesions. In a figure of Rokitansky's handbook (27) star-like intimal cells (the so-called *Bindegewebskörperchen*) are shown. Furthermore, accumulations of small fat droplets are to be seen (Fig. 3), which were interpreted by Rokitansky (27) as contents of the star-like intimal cells. Virchow (42) described globular cells filled with fat droplets. He derived these cells from intimal cells.

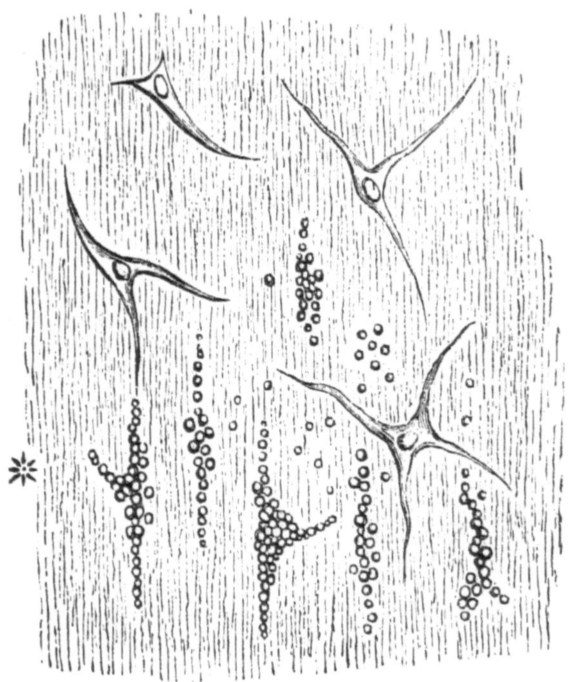

Fig. 3. Star-like intimal cells partly filled with fat droplets. From Rokitansky (27).

Later on, these cells were called Langhans cells to commemorate the paper written by Theodor Langhans (19) in 1866, when he was 27 years old. Six years later he was offered the chair of pathology of the university of Gießen, and after a short time, he went to Bern. Today, his name persists for the giant cells of tuberculosis and in a cell layer of the villi of the placenta.

Langhans (19) described, with great accuracy, star-like cells in the normal intima with interconnection of their processes (Fig. 4) and fat metamorphosis. He delimited these cells from endothelial cells. In arteriosclerotic lesions, he observed round cells with mitoses and transitions to star-like cells. He supposed the star-like cells to be contractile and mobile. Furthermore, Langhans described transitions from smooth muscle cells of the media to intimal cells.

Electron microscopic investigations in the early 1950s allowed a definite classification. Today, we know that modified smooth muscle cells are the cells of the normal intima and of arteriosclerotic plaques in man as well as in experimental animals (12). They are entirely responsible for the formation of connective tissue fibers in arterial walls under normal and pathologic conditions. During vascular development there are, for example, in the media of mammalian arteries no other cells than smooth muscle cells (4).

There is a broad spectrum of phenotypes of smooth muscle cells from almost exclusively contractile cells to modified synthesizing cells. The latter contain much rough endoplasmatic reticulum, ribosomes and well developed Golgistructures, but only a few filament bundles. In arteriosclerotic lesions modified smooth muscle cells synthesize large amounts of extracellular matrix, especially collagen fibres.

In arteriosclerotic plaques histogenetic classification of foam cells was a continuous puzzle. Light microscopic investigations have shown that there are two types of fat-containing cells in the intima: rounded ones and narrow, ramified cells. Today, the term foam cell for these cells is most common. Monocyte-derived macrophages,

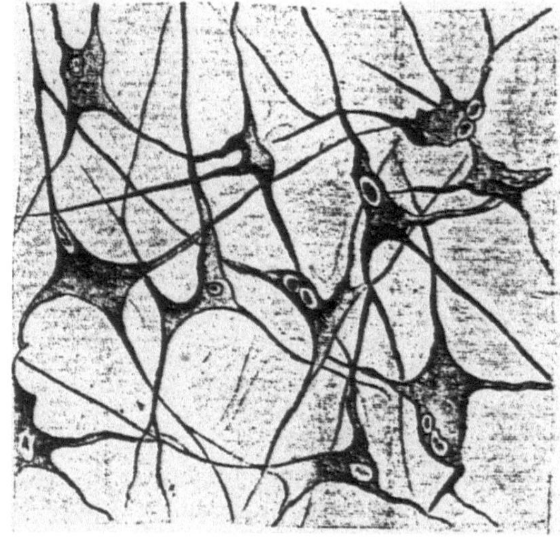

Fig. 4. Star-like cells in the arterial intima. From Langhans (19).

white blood cells, endothelial cells, fibroblasts and smooth muscle cells were supposed to be the histogenetic origin.

In the 1960s, with the help of the electron microscope, many of these fat-containing cells and, especially, the narrow ones were identified as smooth muscle cells. In the following years progression of immunohistochemistry helped to identify other intimal cells as monocyte-derived macrophages. Today, we can say that there is generally a mixed cell population in the intima with varying amounts of smooth muscle cells and monocytes.

Foam cells are the predominant cell type of fatty streaks. These flat lesions may develop into raised fibrous plaques. Frequently, their growth takes place in steps due to further proliferation of intimal cells or by incorporation of mural thrombi. Both mechanisms are possible, but we do not know the frequencies.

With increasing thickness of stenosing plaques, necroses in basal regions of the intima (Fig. 5), atheroma, and accumulations of foam cells arise. Several mechanisms were supposed to induce necroses of arteriosclerotic plaques: local ischemia due to the increasing distance between the vasa vasorum in the outer media and the inner surface of the intima, action of foam cells, increased osmotic pressure in the atheroma or influence of toxic substances or of the tumor necrosis factor.

Main components of atheroma were described by Rokitansky (27) and Virchow (41, 42). They observed crystals of cholesterol (Fig. 6), foam cells and necroses. Furthermore, they knew that rupture of arteriosclerotic plaques is followed by ulceration. Virchow (41) compared the rupture of an arteriosclerotic plaque to the perfora-

Fig. 5. Arteriosclerotic plaque with basal necrosis in a coronary artery. HE stain, 65x.

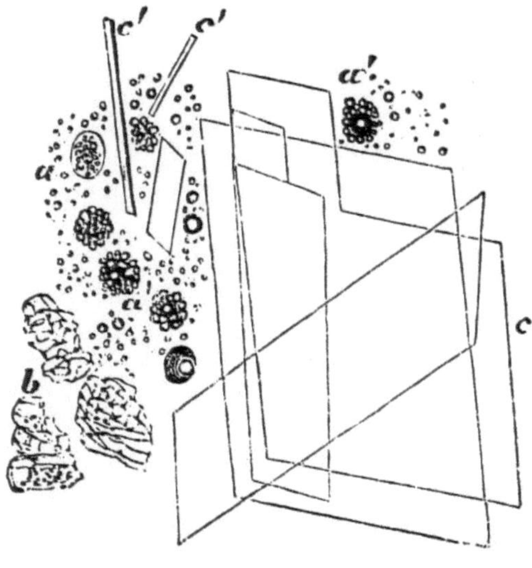

Fig. 6. Crystals of cholesterol (c) and foam cells (a) of an atheromatous plaque. From Virchow (42).

tion of an abscess. The main object of these investigations was the aorta. The coronary arteries were not mentioned, neither by Rokitansky nor by Virchow.

Today, we know that nearly all coronary thrombi arise in the area of the rupture of the fibrous cap of an atheromatous plaque (Fig. 7). But up tonow, we still do not know the exact mechanism leading to the rupture.

At present, the immigration of monocytes into the intima during the early stages of arteriosclerosis is receiving growing attention. Often, it isregarded as an indication of an inflammatory genesis of arteriosclerosis (22, 31, 32).

This idea is not new at all. It was already formulated before Virchow. Virchow (41) considered arteriosclerosis to be an inflammatory process and proposed the term "endoarteriitis chronica deformans sive nodosa". He gave the following criteria of inflammation: irritation and proliferation of tissue and active intracellular absorption of such substances that have reached the intima (so-called cloudy swelling). Over the years the concept of inflammation has changed from time to time. Under the influence of the experiments performed by Cohnheim (6) the microcirculation and emigration of white blood cells were the center of attention. Today, immunohistochemical mechanisms are of importance.

The sclerotic changes in the arterial wall take place in an avascular area. This fact contributed to the decreasing support of the inflammation theory of arteriosclerosis in the time after Virchow.

Conversely, with regard to inflammation it is really surprising that, internationally, one form of arteriosclerosis has hardly received any attention: the early gelatinous lesion. It was known to Virchow (41) and Rokitansky (27), and W. W. Meyer (24) investigated that form thorougly. He thought the insudation to be of utmost importance. This is an inflow of plasma mixed with protein and fibrin directed from the lumen towards the intima. This process is nearly identical with the serous inflammation described by Rössle (28). This is a very mild form of inflammation which can be followed by fibrosis.

7

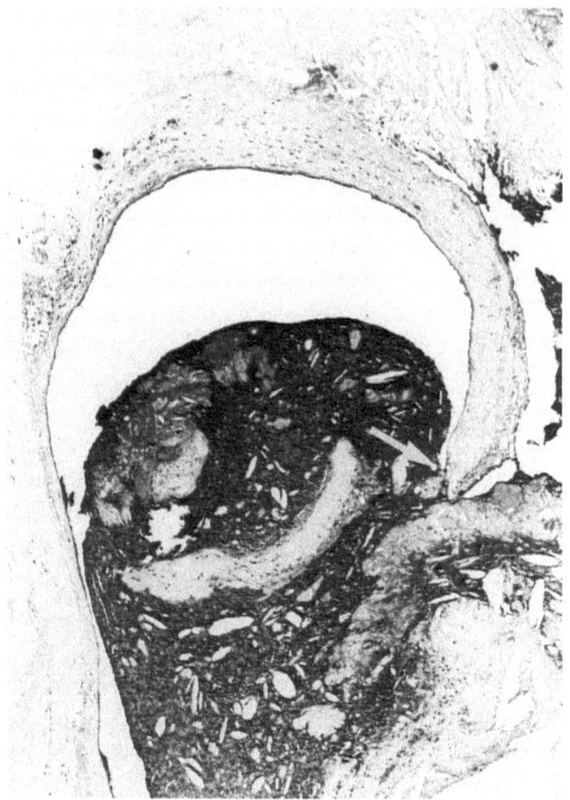

Fig. 7. Rupture (arrow) of an atheromatous plaque with a coronary thrombus and atheromatous debris. HE stain, 65x.

Larger gelatinous plaques in the aorta are relatively frequent, microscopically small ones in the coronary arteries are rare (Fig. 8).

With regard to inflammation, one has to take into account that arteriosclerosis can evolve differently. The vessel wall has limited possibilities of reacting and one component, e.g., immigration of monocytes alone is hardly sufficient to define inflammation.

Arteriosclerosis does not show the whole range of inflammatory changes. Three of four cardinal signs of inflammation (calor, dolor, and rubor) are missing. Perhaps the process is characterized correctly when some early stages of arteriosclerosis are regarded as a borderline-type of inflammation. In later stages degenerative changes are predominant.

Arteriosclerosis is not a uniform disease. Therefore, no generally accepted definition exists. Likewise, the term arteriosclerosis is not perfect. The first part – "artery" – is always correct, but induration arises only in later stages. Today, the term atherosclerosis created by Marchand (23) is mostly preferred. A disadvantage in this term is that, in early stages, both atheroma and induration are missing. Therefore, the term arteriosclerosis seems to be better. Perhaps a triviality contributes to the popularity of the term atherosclerosis: It is easier to pronounce.

8

After this little excursion through the past, let us turn to some contemporary unresolved problems. Here, I would like restrict myself to the localization and distribution of arteriosclerotic lesions.

Feyrter (9), an outstanding pathologist born in Austria, stated that the mystery of localization is a general problem in pathology.

Is the localization of arteriosclerotic lesions still a mystery? Risk factors alone are not helpful to answer this question. For example, in the whole arterial blood the same levels of lipids and – in smokers – of nicotine and carbonmonoxide prevail and the differences of blood pressure are very small in larger arteries.

Probably several factors are involved, for example, vascular geometry, architecture, and environment of arteries and hemodynamics.

Concerning vascular geometry the severity of arteriosclerosis is directly related to the caliber of the vessel, probably due to mural tension (36).

In some regions architectural differences in arterial walls probably correlate with their vulnerability to arteriosclerosis (13).

The importance of the environment of arteries is evident from the reduced severity of arteriosclerosis in the region of muscle bridges. This anomaly is found rather frequently. In nearly a quarter of humans a segment of the left descendent coronary

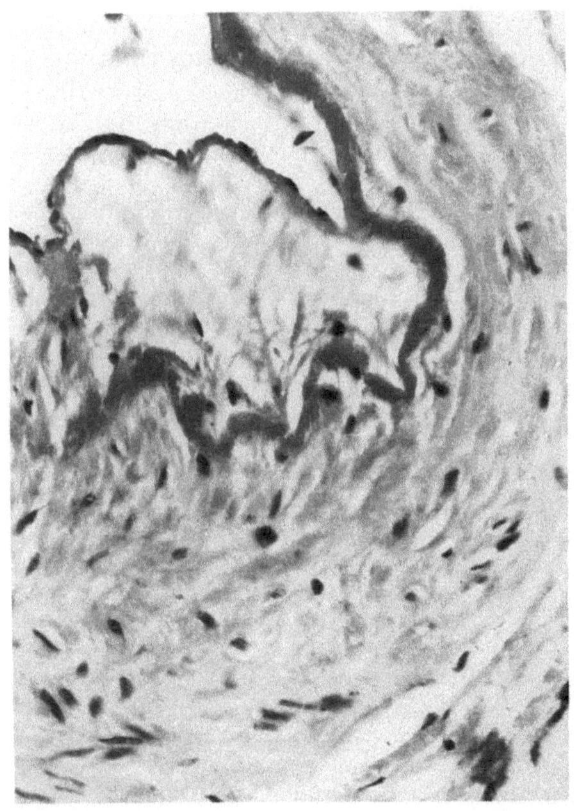

Fig. 8. Circumscribed edema (small gelatinous plaque) in the intima of a coronary artery branch. HE stain, 475x.

artery is exposed to the cyclic contractions of the surrounding heart muscle fibers. One can speculate that coronary heart disease would be rare if coronary arteries in humans were covered by a thin layer of heart muscle cells as, for example, in rats.

Can hemodynamics contribute to the understanding of the topographical distribution of arteriosclerosis? Some findings indeed suggest a relationship between hemodynamic disturbances and localization of atherosclerotic lesions, but detailed physiological information is missing for most vessels, e.g., coronary arteries.

It is an exciting question whether morphology can provide information about hemodynamic conditions or not.

Twenty years ago the group of Donald Fry (10) from the NIH studied the orientation of endothelial nuclei of the middle and lower canine thoracic aorta. Under normal conditions these nuclei are oriented parallel to the long axis of the vessel. Experimentally, the authors removed a segment of the descending thoracic aorta and opened it longitudinally. It was turned around by 90 degrees and reimplanted. Now, in this segment, endothelial nuclei were oriented vertically to the others. Within 10 days after surgery the major axis of the endothelial nuclei was again in agreement with the long axis of the aorta.

These experiments show that nuclei of endothelial cells are sensitive to local hemodynamic stress. It seems that not only the fluid shear stress, but also the stretching and relaxing of the arterial wall (7) are main factors which account for endothelial orientation. From these and other observations the question arose if the orientation of endothelial cells and, especially, of their nuclei could be regarded as an indicator of the properties of the blood flow in the neighborhood.

To prove this concept the sinus of the internal carotid artery is a very suitable model. Here, the characteristics of the blood flow are well known. Formation of arteriosclerotic plaques primarily occurs in the outer part of the sinus where blood flow separation is going along with the development of vortices and low shear stress. On the other hand, in areas with predominantly laminar flow atherosclerotic lesions are rare.

Starting from these findings, K. F. Bürrig and I (2) observed that endothelial nuclei in the outer wall of the carotid sinus were less oriented and more polymorphic than those in the common carotid artery (Fig. 9). We concluded that the disturbed endothelial pattern could be considered as predicting the development of atheromatous plaques (2).

Today, in the coronary arteries the characteristics of the blood flow are not well known. We suspected to find similar disarrays of endothelial pattern at the major sites of coronary arteriosclerotic plaques as in the internal carotid artery. But our results were surprising.

In the normal left descendent coronary artery of 11 human hearts and six monkey hearts (macaca fasciculata), we found a rather monotonous endothelial pattern (3). The nuclei were orientated in the long axis of the vessels, indicating a unidirectional laminar flow. The lack of disturbances of the endothelial pattern suggests a lack of significant flow irregularities. Under normal conditions and normal blood pressure probably factors other than hemodynamic irregularities account for coronary arteriosclerosis. Hitherto, we did not know exactly whether the fixation of the descendent coronary artery on the myocardium (37) or other conditions are essential factors in the natural history of coronary arteriosclerosis.

In opposition to normal coronary arteries, the endothelial pattern in high grade stenosing plaques is changed (1). Upstream of the stenosis, endothelial cells are normally arranged. In the area of severe stenosis they are elongated and their cytoskele-

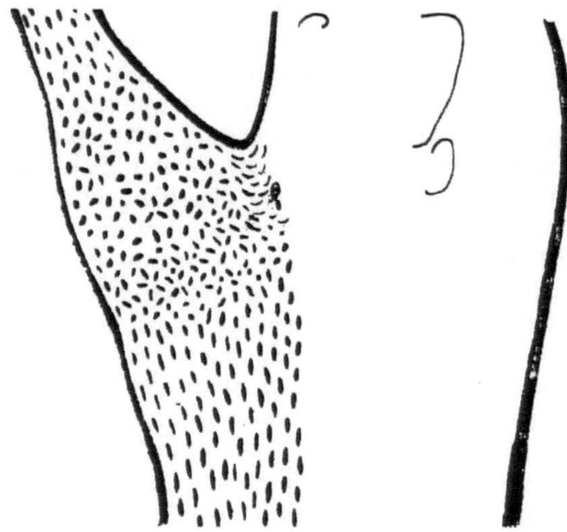

Fig. 9. Scheme of the endothelial pattern at the carotid sinus and the common carotid artery. From Bürrig and Hort (2).

ton is hypertrophied. Most irregularities appear at the outlet of the stenosis. Here, endothelial cells are smaller, with a polygonal cobblestone appearance. This pattern indicates irregularities of blood flow, probably with true turbulences. Probably, these irregularities favor the progression of stenotic lesions. This assumption is in agreement with a centrifugal extension of coronary intimal plaques, described by Velican et al. (39).

Impressed by the dangerous stenosing lesions, we should not forget that arteriosclerotic plaques in humans arise in an intima which shows, even under physiological conditions, a diffuse thickening. This, so to speak, physiological diffuse intimal thickening is, for example, pronounced in the aorta and coronary arteries. It starts in infancy and increases during lifetime. It is more pronounced than in animals (5) and it is interesting that, in comparative pathology, susceptibility to arteriosclerosis correlates with physiologic intimal thickening. In some animal species, for example, in rats, there is a very thin intima of coronary arteries nearly free of muscle cells. Here, in experimental arteriosclerosis, smooth muscle cells of the media are the main source of plaques. But in humans in the 4th decade the diameter of the intima is equal to the diameter of the media (Fig. 10) and the intima already contains a lot of modified smooth muscle cells. But the "physiological" intimal thickening is not uniform. There are, for example, prominences in the left coronary artery opposite to the flow divider (34, 35).

The key role of the endothelium in the development of arteriosclerosis is well founded. It tacitly implies that the inner layer of arteries exclusively consists of endothelial cells. But there are exceptions to the rule. In the aorta of rats nearly 10% of the cells in the endothelial layer are of nonendothelial origin (11) and most of them are monocytes. On the surface of human arteriosclerotic plaques, we observed, as well, cells which appeared to be of non endothelial origin (18). Here, further investigations are necessary to elucidate whether in old age or under pathologic conditions the endothelial layer becomes inhomogeneous. A mixed cell population in the endothelial layer should be followed by functional changes, for example, with modified permeability.

11

To some authors, arteriosclerosis, as a rule, appears to be a systemic disorder with multi-organ involvement. For example, clear associations between ischemic heart disease, carotid stenosis and femoropopliteal disease are described (33, 38, 43). But how significant are these associations? In our investigations, we have quantified the severity and the extent of arteriosclerotic lesions of the descending aorta, internal carotid artery, left descending coronary artery, and femoral artery of 102 autopsy cases (17). Depending on the extent of the lesions, each case was classified in one out of four groups, each group covering an interval of 25 %. Very seldom were all four arteries of an individual case found in the same group, but they were mostly scattered over one or two groups. That means that different arteries of one individual case showed significant differences in the extent of lesions. In nearly 20 % of all investigated cases, however, one artery was found in group 1 showing no or a small lesion extent, whereas another artery was found in group 4 with extreme lesion extent.

In larger data sets there are correlations between the severity of arteriosclerotic lesions in different arteries. But for individual cases, unfortunately, at present a certain prediction is impossible. Our understanding of the mechanisms causing these differences is still very limited. Further investigations on risk factors, hemodynamic pecularities of individual arteries, their structure and environment are needed in order to extend our understanding of arteriosclerosis.

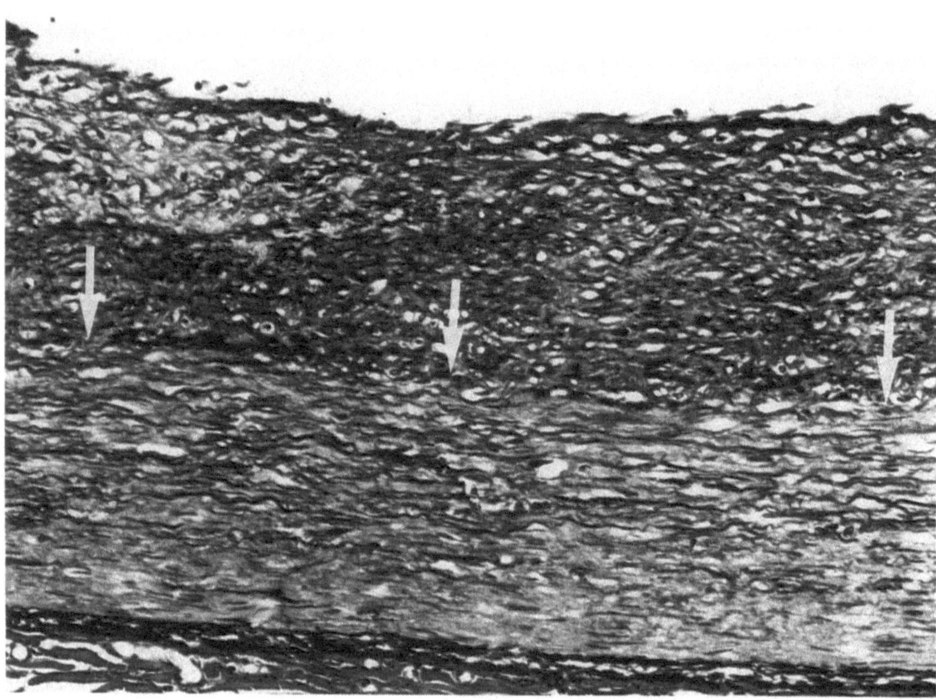

Fig. 10. Coronary artery without arteriosclerotic plaque in an adult man. The thickness of the intima corresponds with the thickness of the media. Arrows: Luminal margin of the media. E-vG stain, 150x.

12

Discussions concerning the severity of arteriosclerosis are mostly confined to the arteries with highest clinical importance: Aorta, cerebral, coronary and femoral arteries. But there are other arteries with severe arteriosclerotic lesions, too.

Investigating many slides of the prostate gland, we were astonished about the severity of arteriosclerosis around the prostate gland and in its capsule. This fact was already known to some pathologists nearly 100 years ago (14, 20, 25). Unfortunately, it was nearly forgotten over recent decades. We picked up this problem and quantified the degree of stenoses of these arteries and found it to be more pronounced than in coronary artery branches of the same size (44).

Up to now, it is, for example, not known whether the severity of periprostatic ateriosclerosis is due to the pronounced tortuosity of supplying arteries. An explanation of this phenomenon could contribute to our understanding of the pathogenesis of arteriosclerosis in general.

Finally, let us return to the beginning of the lecture which I opened with the little anecdote about the astronomer Argelander. During the last few decades there has been a breathtaking progress in natural and medical sciences. For example, new insights were obtained into the origin and development of the stars. We have had the opportunity to experience that, today, reaching out for the stars is more than a dream: it starts becoming reality.

In a comparable manner, today's medicine is able to dilate stenosed coronary arteries, to resolve thrombi in coronary arteries and other blood vessels, and to perform heart transplantations. Insights are gained into the molecular biology of some disturbances. Today, for example, in some forms of hypertrophic obstructive cardiomyopathy the location of the pathologic gene is known (30) and through gentechnologic experiments performed in mice, production of human apolipoprotein A1 was induced which protects against the development of fatty streak lesions (29).

But besides reaching out for "medical stars", we should not forget the small ways and the little, troublesome paths on earth. In view of the spectacular results of curative medicine, we shoud not forget that prevention of arteriosclerosis is much more successful then the efforts to cure the lesions.

In antiquity, for example, in the works of Hippokrates, wide space was devoted to the principles of healthy life and prevention. These ideas still persisted up to the late Middle Ages and guide books to healthy life, the so called "regimina sanitatis", were widespread.

Furthermore, in the Middle Ages prevention was firmly established in medical education. This is known for example from the books kept by the deans of the medical faculty of Cologne, preserved from 1491 up to the 18th century. In these books questions asked in different examinations were collected. Many more of these old questions were concerned with a healthy life and preventive medicine then the ones asked today (16).

At present, the position of preventive medicine in our country is rather insignificant in comparison to the spectacular results of curative medicine. It seems to me that preventive medicine is running the risk to be regarded as a second class science and, today, it is difficult to inspire young and talented scientists to enter this field.

This short historical review has shown that some recent hot topics are not entirely new. Often, it is a pleasure to read old papers because of their abundance of ideas which indicate that some of the old scientists had more ideas, whereas today we have more technology. Some of the old ideas were so fruitful that they were revived, and sometimes we have the opportunity to witness that today's old science evolves into tomorrow's new science.

References

1. Bürrig KF (1991) The endothelium of advanced arteriosclerotic plaques in humans. Arteriosclerosis and Thrombosis 11: 1678–89

2. Bürrig KF, Hort W (1988) Pathogenesis of carotid atherosclerosis. In: Hennerici M, Sitzer G, Weger HD (eds) Carotid artery plaques. Pathogenesis – Development – Evaluation – Treatment. Karger Basel 101–114

3. Bürrig KF, Hort W (1991) The endothelial pattern of the left anterior descending coronary artery in humans and subhuman primates indicates unidirectional blood flow with no significant secondary flow phenomena. Coronary artery disease 2: 493–500

4. Campbell GR, Campbell JH (1987) Smooth muscle cells. In: Olsson AG (ed) Atherosclerosis. Biology and clinical science. Curchill Livingstone Edinburgh 105–115

5. Cucu F (1980) Phylogenetic evolution of the coronary intima and its relevance to atherosclerotic involvement. Med int 18: 99–104

6. Cohnhein J (1867) Über Entzündung und Eiterung. Virchows Arch 40: 1 (Reprint in: Klassiker der Medizin, K. Sudhoff [ed] Barth Leipzig 1914, 10–85)

7. Dartsch PC. Betz E (1989) Response of cultured endothelial cells to mechanical stimulation. Basic Res Cardiol 84: 268–281

8. Doerr W (1987) Die Pathologie Virchow's und die Lehre von der Arteriosklerose. Pathologe 8: 1–8

9. Feyrter F (1931) Zur Geschwulstlehre (nach Untersuchungen am menschlichen Darm) I. Polypen und Krebs. Beitr. Path. Anat. 86: 663–760

10. Flaherty JT, Pierce JE, Ferrans VJ, Patel DJ, Tucker WK, Fry DL (1972) Endothelial nuclear patterns in the canine arterial tree with particular reference to hemodynamic events. Circ Res 30: 23–33

11. Freudenberg N, Riese KH, Freudenberg MA (1983) The vascular endothelial system. Fischer Stuttgart

12. Geer JC, Haust MD (1972) Smooth muscle cells in atherosclerosis. Monographs on Atherosclerosis Vol 2. Karger Basel

13. Glagov S (1972) Hemodynamic risk factors: Mechanical stress, mural architecture, medial nutrition, and the vulnerability of arteries to atherosclerosis. In: Wissler RW, Geer JC The pathogenesis of atherosclerosis. Williams and Wilkins, Baltimore, p 164

14. Guyon F (1888) La vessie et la prostate. Lecons cliniques sur les affections chirurgicales de la vessie et de la prostate. Baillere et fils Paris

15. Henle J, see (8)

16. Hort I: Personal communication

17. Hort W: To be published

18. Hort W, Bürrig KF (1989) Endothel und Arteriosklerose. Z. Kardiol 78 Suppl 6: 105–112

19. Langhans T (1866) Beiträge zur normalen und pathologischen Anatomie der Arterien. Arch path Anat Physiol 36: 187–226

20. Launois P (1885) De l'appareil urinaire des vieillards. Steinheil, Paris

21. Lobstein JE (1833) Traité d'anatomie pathologique. Tome II. Levrault Paris

22. Majno G, Joris I, Zand T (1985) Atherosclerosis: New horizons. Hum Pathol 16: 3–5

23. Marchand F (1904) Über Arteriosklerose (Athero-Sklerose). Verh Kongr Inn Med 21: 23

24. Meyer WW (1949) Die Bedeutung der Eiweißablagerungen in der Histogenese arteriosklerotischer Intimaveränderungen der Aorta. Virchows Arch 316: 268–316

25. Oberndorfer S (1931) Die inneren männlichen Geschlechtsorgane. In: Henke F, Lubarsch O (eds) Handbuch der speziellen Pathologischen Anatomie und Histologie Bd 6, 3. Teil. Springer, Berlin

26. Rokitansky C (1844) Handbuch der speciellen pathologischen Anatomie, Bd I. Braumüller und Seidel, Wien

27. Rokitansky C (1855) Lehrbuch der Pathologischen Anatomie. 3. umgearbeitete Aufl., Bd I. Braumüller und Seidel, Wien

28. Rössle R (1944) Über die serösen Entzündungen der Organe. Virchow Arch 311: 252–284

29. Rubin EM, Krauss RM, Spangler EA, Verstuyft JG, Clift SM (1991) Inhibition of early atherogenesis in transgenic mice by human apolipoprotein A1. Nature 353: 265–267

30. Seidman CE, Seidman JG (1992) Mutations in cardiac heavy-chain genes cause familial hypertrophic cardiomyopathy. Bas Res Cardiol 87 Suppl 1: 175–185

31. Schwartz CJ, Valente AJ, Sprague EA, Kelley JL, Suenram CA, Rozek MM (1985) Atherosclerosis as an inflammatory process. Ann New York Acad Sci 454: 115–120

32. Schwartz CJ, Valente AJ, Sprague EA, Kelley JL, Suenram CA, Graves DT, Rozek MM, Edwards EH, Delgado R (1986) Monocytemacrophage participation in atherogenesis. Sem thromb haemost 12: 79–86

33. Stavenow L, Karlsson S, Lilja B, Lindgärde F (1988) High prevalence of coronary heart disease in patients with intermittent claudication. Acta Chir Scand 154: 447–451

34. Stary HC (1983) Structure and ultrastructure of the coronary artery intima in children and young adults up to age 29. In: Atherosclerosis VI, Schettler FG et al (eds), Springer Berlin, pp 82–86

35. Stary HC, Blankenhorn DH, Chandler AB, Glagov S et al (1992) A definition of the intima of human arteries and of its athesclerosis-prone regions. A report from the comittee on vascular lesions of the council on arteriosclerosis, American Heart Association. Circulation 85: 391–405

36. Stehbens WE Localization of atherosclerotic lesions in relation to haemodynamics. In: Atherosclerosis. Biology and clinical science Olsson AG (ed) Churchill Livingstone Edinburgh, 175–182

37. Strunk W, Bürrig KF, Hort W (1990) Quantitative morphologische Untersuchungen zur Lokalisation arteriosklerotischer Polster in der Zirkumferenz der Kranzarterien. Z. Kardiol 79: 273–278

38. Travers AM, Nel CJC, Barry R, Pienaar CW, Filmater B (1990) Atherosclerosis – multiorgan involvement the rule rather than the exception. S Afr Med J 77: 140–143

39. Velican C, Velican D, Tancu I (1986) Centrifugal extension of coronary intimal necrotic areas. Rev Roum Med. Med Int 24: 93–101

40. Virchow R (1846) Preußische Medicinal-Zeitung XV: 237 und 243

41. Virchow R (1856) Gesammelte Abhandlungen zur wissenschaftlichen Medicin. Meidinger, Frankfurt

42. Virchow R (1858) Die Cellularpathologie in ihrer Begründung auf physiologische und pathologische Gewerbelehre. 1. Aufl. Hirschwald, Berlin

43. Witteman JCM, Kannel WB, Wolf PA, Grobbee DE, Hofman A, D'Agostino RB, Cobb JC (1990) Aortic calcified plaques and cardiovascular disease (The framingham study) Am J Cardiol 66: 1060–1064

44. Wolf M (1992) Morphometrische Untersuchungen an periprostatischen Arterien. Med Diss Düsseldorf

Author's address:
Prof. Dr. W. Hort
Pathologisches Institut
der Heinrich-Heine-Universität
Moorenstraße 5
D-40225 Düsseldorf
FRG

Changes in components and structure of atherosclerotic lesions developing from childhood to middle age in coronary arteries

H. C. Stary

Louisiana State University School of Medicine, New Orleans, USA

Summary: The composition and structure of adaptive intimal thickening and of atherosclerotic lesions that can develop in human coronary arteries is described. Adaptive thickening occurs in defined locations from birth and represents a self-limited response of the intima to hemodynamic forces present within specific locations. Adaptive thickening does not indicate or presage an atherosclerotic lesion. However, some of the identical intima locations (progression-prone locations) accumulate more lipoprotein in persons exposed to risk factors of athero-sclerosis and are first to develop advanced lesions if such lesions develop at all.

Atherosclerotic disease can be resolved into eight (I-VIII) lesion types, each characteristic by its cells, matrix, architecture, or other specific features. The numerals I-VI represent the usual sequence in which lesions develop and progress from the initial accumulations of lipopro-teins and macrophages to atheroma and fibroatheroma stages which are susceptible to thrombo-tic deposits and ischemic clinical episodes. The numerals VII and VIII represent morphological variants that may follow or precede Type VI. Types I-IV are the lesions most frequent in the first four decades of life. Type III is a lesion we identified in adolescents and young adults as mor-phologically intermediate between the small lesions of children (I and II) and the potentially symptom-producing Type IV lesion.

Identification of Type III provides evidence that small lesions of children can develop into clinical ones. Because we know the age at which Type III lesions are present in our population, we also know the age when progression to advanced lesions generally begins and when preven-tive measures should already be in place.

Key words: Atherogenesis – smooth muscle phenotypes – macrophages – extracellular lipid – calcification – thrombosis

Introduction and methods

Accumulation of lipid in the intima and associated reactions begin in childhood and adolescence in the majority of our population and, when continued or increased, may lead to symptom-producing atherosclerotic lesions at middle age or later. The mechanisms involved in this long process are not entirely clear. Once symptom-pro-ducing lesions are present, lesion composition and architecture are complex and pathogenetic mechanisms are difficult to unravel and decipher. Atherogenesis can be better understood, and more successfully influenced, before lesions become clini-cally overt. In the studies summarized in this article we have tried to clarify the sequence in which intimal cell types and interstitial matrix components change and new ones appear as lesions develop in childhood or youth and as they proceed to clin-ical disease. Lesion components addressed specifically include intimal smooth mus-cle cell phenotypes, macrophages, neovascularization, collagen formation, calcifica-tion, hemorrhage, and thrombosis.

"

From 1979 to 1988, we obtained coronary arteries and aortas from 1286 persons who died between birth and the age of 39 years. We studied these vessels and their atherosclerotic lesions by various techniques. Since we could not examine a specific intima location or an initial lesion and then follow its behavior over a lifetime, we determined instead the nature of the intima and of initial changes in precisely defined arterial locations in infants and children and then studied the same anatomic locations in adolescents and young and middle-aged adults. The arterial locations chosen for particular study were those known for their predisposition to develop obstructive lesions in adults. We examined enough lesions at each age to satisfy ourselves that the complete evolutionary sequence as it occurs in most people with atherosclerosis is represented in our sample.

Most of the 1286 subjects died of accidental causes or violence but were otherwise apparently healthy. Because the time between death and autopsy was relatively short in 691 in the subjects (mean interval 9.5 h), we distended and fixed their coronary arteries by perfusing them with glutaraldehyde under physiologic pressure. All 691 cases were studied by high-resolution light microscopy and morphometric techniques and a subgroup was studied by electron microscopy. Routinely, six one-micron thick cross-sections spaced at more or less equal intervals were prepared from a precisely defined coronary segment extending for a length of up to 25 mm. To ascertain the three-dimensional composition, we cut up to 25 one-micron-thick cross-sections of the entire extent of a lesion in a subgroup of cases. From the 691 cases a total of about 6000 one-micron thick coronary artery cross-sections (with or without lesions) were examined by light microscopy and the results summarized in this paper are based on their evaluations. Particulars of the methods used to collect, process, and evaluate the coronary arteries can be found in earlier publications (30–32). Lesions in the aortas of the 691 cases were also studied in great detail. The methods used to study the aortas and the data obtained will be reported at a later date.

Adaptive increases in intimal thickness: eccentric and diffuse intimal thickening

The arterial intima of humans is normally of unequal thickness. Regions of relatively thick intima are present from infancy, develop apparently in the fetus, are self-limited in growth, and do not obstruct blood flow at any age (33). The thick regions represent physiological adaptations to specific mechanical forces, secondary to local changes in blood flow or wall tension within an artery. In these locations wall shear stress is reduced or wall tensile stress is elevated or both may be the case (5, 39). Adaptive thickening is defined here to allow its separation from atherosclerotic and other vascular disease. The terms *eccentric* and *diffuse* are used to differentiate between two patterns although the two may be contiguous and sometimes difficult to delineate.

Eccentric thickening is a focal increase in the thickness of the intima associated with branches and orifices. At an arterial bifurcation, the thickening involves about half the circumference opposite a flow divider and extends for a short distance along the length of the parent and daughter vessels. In a cross-section of an artery fixed under physiological pressure, eccentric thickening is a crescent-shaped increase in intimal thickness. At the thickest point of the crescent, intima may be up to twice the thickness of the media in coronary arteries of children, although considerable individual variation in degree has been found (30). Eccentric thickening has been seen in the aorta and in coronary, carotid, cerebral, and renal arteries. Its three-dimensional extent and thickness have been graphically outlined in the left coronary artery (31).

Diffuse intimal thickening is a spread-out and often circumferential pattern of adaptive intimal thickening not clearly related to specific geometric configurations of arteries. In coronary arteries the degree of diffuse thickening is less than that of eccentric thickening, although more extensive.

Adaptive intimal thickening is composed of two layers although the two layers may not always be clearly demarcated, especially when the thickening is minimal. The inner layer, subjacent to the lumen, has been called the *proteoglycan layer* because it contains abundant nonfibrous connective tissue identified as proteoglycan ground substance by electron microscopy (22, 36). Elastic fibers are scarce. Smooth muscle cells are both RER-rich and myofilament-rich phenotypes and occur as widely spaced single cells rather than in layers. The part of the proteoglycan layer near the endothelium contains isolated macrophages. The thicker layer underlying the proteoglycan layer (and adjacent to the media) is the *musculoelastic layer* because of an abundance of smooth muscle cells and elastic fibers. This layer also contains more collagen. Smooth muscle cells are of the myofilament-rich phenotype and arranged in close layers.

Regions of the intima with adaptive thickening differ functionally from adjacent regions without the thickening. The turnover of endothelial cells (29, 37), smooth muscle cells (29), and the concentrations of lipoproteins (apo B) (24, 28) and other plasma components are greater in adaptive thickening. These increases should not be considered abnormal unless they enter a range associated with tissue damage.

The relationship between adaptive intimal thickening and atherosclerotic lesions

Because adaptive intimal thickenings can impress by their thickness when viewed under the microscope and because they project into the arterial lumen when arteries are studied in their collapsed and contracted postmortem state, they have been designated as atherosclerosis by many authors. Sometimes they have been misinterpreted as arterial stenoses or occlusions.

Smooth muscle cell accumulations can be produced in the arterial intima of laboratory animals by a wide range of artificial exogenous impulses, including endothelial cell denudation through mechanical injury (3, 4, 35). However, the experimental smooth muscle cell accumulations resemble adaptive thickening only superficially. The hypothesis that adaptive intimal thickening should be considered as arterial disease driven by some injurious factor until it becomes symptom-producing is not supported by our observations in young people.

Nevertheless, there is a relationship between adaptive intimal thickening and atherosclerosis. Atherogenic lipoproteins, when excessive in the plasma, tend to accumulate above all in locations in which adaptive thickening is also present. The term *atherosclerosis-prone* has been applied to these intima locations. In fact, some adaptive thickenings tend to accumulate more lipid than others. The term *progression-prone* is applied to this subgroup (see subsequent section on the *progression-prone* Type II lesion). The view that adaptive thickening is an atherosclerotic process has been based in part on this colocalization with prominent lipid accumulation. However, if adaptive thickening is accepted as a self-limited physiological response to hemodynamic forces in specific anatomic locations, then the development of a lesion refers only to changes that are superimposed. The specific hemodynamic forces in these locations cause the thickening whether high concentrations of atherogenic lipoproteins are present or not. When atherogenic plasma lipoproteins exceed critical levels, the same mechanical forces enhance their deposition in adaptive thickening. Cell reactions associated with the accumulated lipid may eventually transform adaptive thickening into a lesion.

General comments on atherosclerotic lesion types

In the following sections, the compositions of eight morphologically characteristic types of atherosclerotic lesions are described. In the first three decades, the composition of lesions (Types I to IV) is relatively predictable, relating primarily to lipid accumulation. In the fourth decade, the composition of advanced lesions becomes less predictable because some begin to increase by mechanisms different from, and additional to those related to lipid accumulation. The latter mechanisms do not occur automatically or in everyone with hyperlipidemia. They are episodic, while, in comparison, lipid accumulation may be more linear.

Type I and II lesions generally are the only ones found in infants and children although they occur in adults also. Such lesions do not thicken the arterial wall appreciably and therefore do not obstruct or modify blood flow. Type III lesions may evolve soon after puberty and, in their composition, are intermediate between fatty streaks (Type II) and atheroma (Type IV). In this paper, the term *advanced lesion* is used as an umbrella term for lesions beyond Type III. Dissolution and disorganization of the intima in a part of an artery is our biological measure to indicate that a lesion has become advanced. *Advanced* in this sense does not necessarily indicate that a lesion is angiographically visible or clinically overt. The numerals I-VI represent the usual sequence in which lesions develop. Advanced lesions of Type IV are relatively frequent from the third decade and Types V and VI from the fourth decade. The numerals VII and VIII represent morphological variants that may follow or precede Type VI, but they generally are found after the first four decades (data on Types VII and VIII are from other studies).

In Type IV, dissolution of intimal architecture is caused by a mass of extracellular lipid (the lipid core). Type V also contains a lipid core, but in contrast to Type IV, the region surrounding the core, and particularly the region above, is thickened and remodelled by layers of collagen. Progression of Types IV and V to greater stenosis and to clinically overt disease is accelerated by thrombotic deposits and/or lesion fissure with hematoma (Type VI lesions). The term *fibrotic lesion* (Type VIII) is applied when dense layers of collagen but little or no lipid are present. Such lesions may represent the end result of lesions that earlier were more typically atherosclerotic or thrombotic. The same is true for *calcific* (Type VII) lesions in which deposits of calcium phosphate and apatite are the predominant components.

Type IV lesions impede flow mildly or not at all and therefore are clinically silent. Type VI are unstable and often occlusive and symptomatic. Types V, VII and VIII may be silent or overt depending on the degree of stenosis they cause. The characteristic compositions of the eight lesion types are summarized in Table 1.

20

Table 1. Nomenclature[1], sequence[2], and descriptions of human atherosclerotic lesions

Recommended terms	Description
Type I (initial lesion)	Lipoprotein accumulation in intima; lipid in macrophages; these changes discernible only microscopically or chemically; no intima disorganization
Type II (fatty streak) IIa (progression-prone: colocalized with specific adaptive thickening) IIb (progression-resistant)	Lipoprotein accumulation in intima; lipid in macrophages and smooth muscle cells; quantities large enough to be visible to the unaided eye but still no intima disorganization
Type III (preatheroma)	All type IIa changes plus multiple deposits of pooled extracellular lipid; microscopic evidence of tissue damage and disorder
Type IV (atheroma)	All type IIa changes plus confluent mass of extracellular lipid (lipid core) with massive structural damage to intima
Type V (fibroatheroma)	All type IV changes plus development of marked collagen layers and smooth muscle cell increase above lipid core
Type VI (complicated lesion) VIa (fissure) VIb (hematoma) VIc (thrombus)	All type IV or V changes plus a thrombotic deposit, and/or hematoma, and/or erosion or fissure
Type VII (calcific lesion)	Any advanced lesion type composed predominantly of calcium; substantial structural deformity
Type VIII (fibrotic lesion)	Any advanced lesion type composed predominantly of collagen; lipid may be absent

[1] Type I and II lesions are sometimes combined as "early lesions", and type IV to VIII as "advanced lesions"

[2] A developmental sequence of lesions I to VI is usual but not inevitable; the numerals VII and VIII denote lesion types rather than a sequence and such lesions are rare in the first four decades of life

Type I lesion (initial lesion)

Initial lesions represent only microscopically and chemically perceivable lipid deposits and accompanying cell reactions in the intima. The histological change is minimal, consisting of isolated groups of single macrophages distended with lipid droplets (macrophage foam cells).

Macrophages without lipid droplet inclusions are twice the number present in intima normally. Initial lesions coinciding with some adaptive intimal thickenings contain more macrophages and macrophage foam cells than initial lesions located outside these thickenings. Intimal smooth muscle cells are without lipid droplet inclusions and their number and phenotypic range are similar to what is found in identical intima segments of children without foam cells (30). Particles of extracellular lipid and cell debris of the type visible by electron microscopy in more advanced lesions are not present although increased lipoproteins are detectable with other methods.

The initial intimal macrophage foam cells appear to be a sequel and a cellular marker of unphysiological inceases in intimal lipoproteins. In rabbits, a high-cholesterol diet fed for 4–16 days caused increased low density lipoproteins in intima of the aorta before macrophage foam cells appeared (25). Experimentally induced accumulations of lipoproteins, macrophages, and macrophage foam cells are most marked in regions with adaptive intimal thickening.

Type II lesion (fatty streak)

Type II lesions are composed of more lipid-laden cells than initial lesions. One or more layers of macrophage foam cells, frequently interdigitating by means of their microvilli, are present. The number of macrophages without droplet inclusions is 1.7 times the number in normal intima (32). Isolated lymphocytes (13, 21) and mast cells may be present, but both are less numerous than macrophages.

As in initial lesions, the number of smooth muscle cells in a segment of the intima with a fatty streak is similar to the number in the same segment of a child without lesions. However, a variable proportion of these smooth muscle cells now also contains lipid droplet inclusions, although generally the number of lipid droplets per smooth muscle cell is smaller than the number per macrophage foam cell. An increase in the proportion of RER and SER, observed within some of both intimal smooth muscle cell phenotypes, is probably a sign of degradation of the intracellular lipid droplets. While most of the electron microscopically visible lipid is within cells, some is extracellular. Fatty streaks differ from more advanced lesions by the very small amount of this type of extracellular lipid and debris, and the absence of visible damage of intimal structure, reparative tissue reaction, or deformity.

Intimal macrophage foam cells tend to accumulate in the lower part of the proteoglycan layer. In most arterial locations the proteoglycan layer is not deep and a few layers of foam cells fill the layer to the level of the endothelial cells. The term *fatty streak* is derived from the layers of foam cells visible with the unaided eye through the endothelial surface. However, the proteoglycan layer is deep in many adaptive intimal thickenings and in this location macrophage foam cells may not be visible from the intima surface.

Type IIa (progression-prone) and IIb (progression-resistant) lesions

The locations in the arterial tree in which Type II lesions develop are relatively constant (8), and have been called *atherosclerosis-prone* (or lesion-prone) locations. However, of the many Type II lesions generally present in a person with average levels if atherogenic lipoproteins only a subgroup will readily proceed to advanced lesions if advanced lesions develop in that person at all. This subgroup is colocalized with specific adaptive intimal thickenings and may be called *progression-prone* or Type IIa. Locations of the arterial tree with IIa lesions are highly predictable and may be called *progression-prone* locations. The larger subgroup of Type II lesions that do not progress, or only slowly, or only in persons with very high plasma levels of atherogenic lipoproteins, may be called *progression-resistant* or Type IIb. Locations in which advanced lesions do not develop readily may be called *progression-resistant* locations.

Type IIa lesions differ morphologically from Type IIb by containing more lipid, macrophages, and mast cells, and probably more lymphocytes. The location of macrophage foam cells and extracellular lipid is characteristic and determined by the colocalized adaptive intimal thickening. While macrophages without fat droplets are most numerous near the endothelial surface, macrophage foam cells accumulate at the bottom of the proteoglycan layer which in an adaptive thickening can be relatively deep. Although, with time, more macrophage foam cells pile up, they may not back up to the endothelial surface. The terms *submerged fatty streak* and *concealed fatty streak* have been applied to this morphological picture. Because of the presence of the many layers of smooth muscle cells of the colocalized adaptive intimal thickening, Type IIa lesions have often been misinterpreted as atherosclerotic disease more advanced than is actually the case.

Type III lesion (preatheroma)

Morphologically, the Type III lesion is the connecting link between a IIa lesion and the first type which we designate as advanced – Type IV (atheroma). Type III lesions are characterized by microscopically visible accumulations of particles of lipid and cell debris to the extent that pools of this material form among the layers of smooth muscle cells of the colocalized adaptive thickening (Fig. 1). The lipid pools replace intercellular matrix and drive smooth muscle cells apart. Neither death nor proliferation of smooth muscle cells are evident at this stage of progression. By this definition, multiple scattered pools of extracellular lipid, disrupting the coherence of some structural intimal smooth muscle cells constitute progression beyond a Type IIa lesion. The distribution of macrophages and macrophage foam cells is the same as in Type IIa lesions. A massive, confluent, accumulation of extracellular lipid (a lipid core – the hallmark of a Type IV lesion) has not yet developed. When human atherosclerotic lesions were studied by lipid physical biochemistry, a lesion connecting fatty streaks and atheroma also became apparent (14, 26). Histologically, the biochemically distinct lesions resembled the preatheroma described in this section (26).

The supposition that symptom-producing lesions have their roots in fatty streaks has been controversial (15, 16, 19, 27). Several reasons account for this skepticism. As fatty streak and advanced lesions had been traditionally viewed, they differed too sharply from each other. There was thought to be a lack of a precise topographic cor-

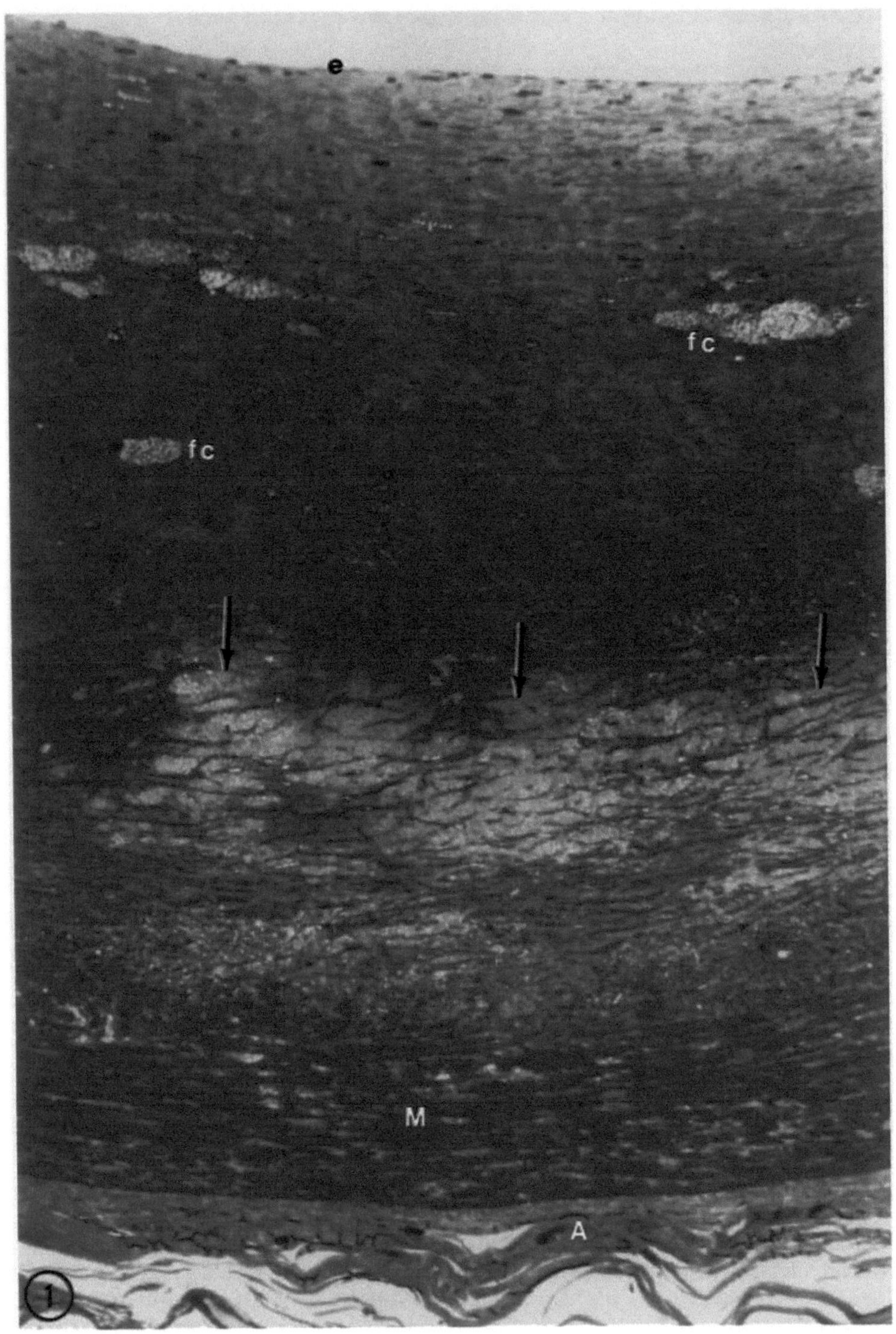

24

respondence between the two lesion types (19). By chemical analysis, the cholesteryl esters of advanced human lesions were found to contain a high proportion of linoleic acid and a low proportion of oleic acid (27), while in fatty streaks the reverse was true (11, 27). If advanced lesions develop from fatty streaks, then a lesion type histologically and chemically intermediate between the two should exist. Although some investigators have suspected and discussed the existence of an intermediate (transitional) morphology (17, 18, 34), this type of lesion has not been previously defined and is not included in past classifications of atherosclerosis. The supposed morphological incompatibility between fatty streaks – as they have been conventionally viewed in thin intima segments – and atheroma is resolved by understanding that fatty streaks developing in youth in progression-prone locations (certain eccentric thickenings) are very rich in macrophage foam cells and extracellular lipid and that, not much later in life, lesions of preatheroma or atheroma morphology are found in the same locations.

Type IV lesion (atheroma)

A Type IV lesion is characterized by a massive aggregate of lipid producing the classic picture of a lipid core within an eccentric adaptive thickening with which it is generally colocalized. The accumulated extracellular particles and droplets have damaged and disorganized intima by displacing structural intimal smooth muscle cells. Smooth muscle cells persisting among the accumulated extracellular material have changed their morphology: they are attenuated and some have thick basement membranes (BM-rich smooth muscle cells).

Lesions categorized as atheroma lack marked smooth muscle cell proliferation, thick layers of collagen, thrombosis, or hematoma. The layer above the lipid core should not, at this stage, be considered as having been entirely formed by the atherosclerotic process. Rather, the layer represents the upper part of the preexisting adaptive thickening which contains macrophages, macrophage foam cells, RER-rich smooth muscle cells with and without lipid droplet inclusions, lymphocytes, plasma cells, mast cells, and capillaries. The periphery (shoulder) of atheroma contains more proteoglycan matrix and macrophage foam cells and fewer smooth muscle cells and is the region most susceptible to fissuring (see Type VI). Lipid cores are constituted of the remnants of many generations of disintegrated macrophage foam cells, incompletely digested lipid droplets (tertiary lysosomes) extruded from smooth muscle cells and macrophages (23), and coalesced lipoprotein particles not previously ingested by cells (12).

◁

Fig. 1. Midportion of a Type III (preatheroma) lesion in the left main coronary artery proximal to the main bifurcation. Extracellular lipid (arrows) is abundant and impinges on some smooth muscle cells of the musculoelastic (deep) layer of an eccentric adaptive intimal thickening. Macrophage foam cells (fc) and lipid-laden smooth muscle cells are above the extracellular aggregates. M = media, A = adventitia, e = endothelial cells at the surface. From a 25-year old man who died in a motorcycle accident. Case no. 372 (P-1372); 1-micron section stained with toluidine blue and basic fuchsin; about x200.

As atheroma exists in most young adults, its lipid accumulation and reactive changes are limited to the intimal thickening and do not at first extend into the media. Nor is the adventitia diseased. Since it coincides with an adaptive thickening of the eccentric type, atheroma is an eccentric rather than concentric lesion.

At this stage, arterial wall thickening is caused mainly by the lipid core which is generally large enough to be visible with the unaided eye. However, measurements indicate that while the wall can be quite thick, arterial lumen is not much reduced, and this first type of biologically advanced lesion may not be visible angiographically. Nevertheless, lesions of the atheroma type can become symptomatic in severe forms of familial hypercholesterolemia through their great lipidic bulk.

Type V lesion (fibroatheroma)

A Type V lesion has formed when, in addition to the components of an atheroma, layers of newly proliferated RER-rich smooth muscle cells and thick layers of collagen are added to and change the nature of the region between the lipid core and the endothelial cell layer at the arterial lumen. The new components sometimes add several fold more thickness to the arterial wall than the underlying lipid core. Granulation tissue including capillaries at the margins of the lipid core may be more abundant than in the atheroma stage and microhemorrhages may be present around the capillaries.

In some cases, the new smooth muscle cell and collagen layers may be the result of thrombotic deposits that became incorporated into the intima. Such changes would have been classified as Type VI if the ingredients of a thrombus could be reliably demonstrated. However, in the Type V lesions of most young adults, increased smooth muscle cells and collagen appear to form gradually and modestly in response to tissue injury and deformity caused by the increasing lipid core alone. Such uncomplicated fibroatheroma does not obstruct the lumen significantly.

Changes in the media and adventitia may be present adjacent to the intimal changes in larger fibroatheroma. At first, islands or columns composed of small blood vessels, loose connective tissue, macrophages and foam cells are found in the media. Media near the intimal lipid core may accumulate extracellular lipid. In larger (more advanced) fibroatheroma, media may become part of the fibrous or fibrocalcific component of the overall lesion. The small blood vessels of the adjacent adventitia may be surrounded by macrophages, macrophage foam cells and lymphocytes.

Fig. 2. Upper part of a Type VIc lesion in the left anterior descending coronary artery about 10 mm beyond its origin at the main bifurcation. A thrombotic deposit (platelets and fibrin) (arrows) is below the surface of a lesion that would be classified Type V if the deposit were not present. Endothelial cells and smooth muscle cells cover the thrombotic deposit. Macrophage foam cells (fc) and lipid-laden smooth muscle cells are above the core of extracellular lipid (core). Macrophages without fat droplets (m) are closer to the endothelial surface (e). From a 30-year old man. Homicide was the cause of death. Case no. 1157 (P-2157); 1-micron section stained with toluidine blue and basic fuchsin; about x220.

fc
m
m
e
core
2

Type VI lesion (complicated lesion)

Lesions classified as Type VI include demonstrable thrombotic deposits and/or hematoma and/or fissure. Generally, these occur as complications of lesions of the atheroma or fibroatheroma type. The result is acceleration in growth of the lesion and stenosis or occlusion of medium-sized vessels such as the coronary arteries. Type VI may be subdivided to reflect the dominant complicating factors: VIa to indicate fissure, VIb hematoma, and VIc thrombus (Fig. 2). Lesions designated VIabc contain all three components.

Thrombotic deposits may be visible only with the microscope, or they may be large enough to be visible with the unaided eye. Clinically (angiographically), evidence of complication connotes lesion instability and the risk of precipitous changes leading to ischemic clinical episodes. Type VI lesions are thought of as the morphological counterpart of unstable angina (10). Thrombotic deposits accelerate lesion growth by stimulating intimal smooth muscle proliferation and collagen production. When thrombotic deposits are visible on the surface or within lesions, then the number of smooth muscle cells as well as the amount of collagen and the thickness of the lesions is more often greater than when such evidence is not present.

The causes of hematomas and/or thrombotic deposits are multiple. Erosion or ulceration of the lesion surface have long been known to constitute one cause. Some authors published evidence that shearing tears (fissures) of the lesion surface are a principal cause of massive hemorrhage into the lesion, thrombotic deposits, rapid lesion expansion, and symptomatic disease (6, 7, 9). Although large hematomas appear to be the consequence of shearing tears or erosions of the surface, there is evidence that smaller hemorrhages within lesions are caused by breaks in newly formed capillaries (1, 2). Even the Type IV and V lesions of young adults often contain capillaries, and microhemorrhages may surround them. These are too small to cause breaks in the lesion surface or thrombosis and we have not classified lesions with microhemorrhages as Type VI.

Thrombotic deposits can form on lesions without a surface defect or hemorrhage. The causes may include changes in blood flow secondary to deformity of the surface by the underlying lesion, facilitating platelet deposition particularly in persons with some hypercoagulable states. High plasma fibrinogen levels have been found in persons with clinical ischemic episodes (20, 38). Functional impairment of endothelial cells or loss of small groups of endothelial cells, impossible to detect even microscopically, might also facilitate thrombus formation.

Figure 3 indicates that Type VI lesions are atheroma or fibroatheroma lesions in which a hematoma and/or a thrombotic deposit have developed. Tears, hematoma, and thrombotic deposits may be replaced with collagen completely and the lesion then appears, and is relabeled, as Type V, although a thicker lesion now than before its temporary sojourn as Type VI. Thus, the numeral VI indicates the present (momentary) composition of the lesion and not necessarily a greater thickness or a greater degree of vascular narrowing than Type V.

Type VII (calcific) lesion

Some advanced atherosclerotic lesions, particularly after the fourth decade, are largely mineralized. The term *calcific lesion* (Type VII) may be applied here.

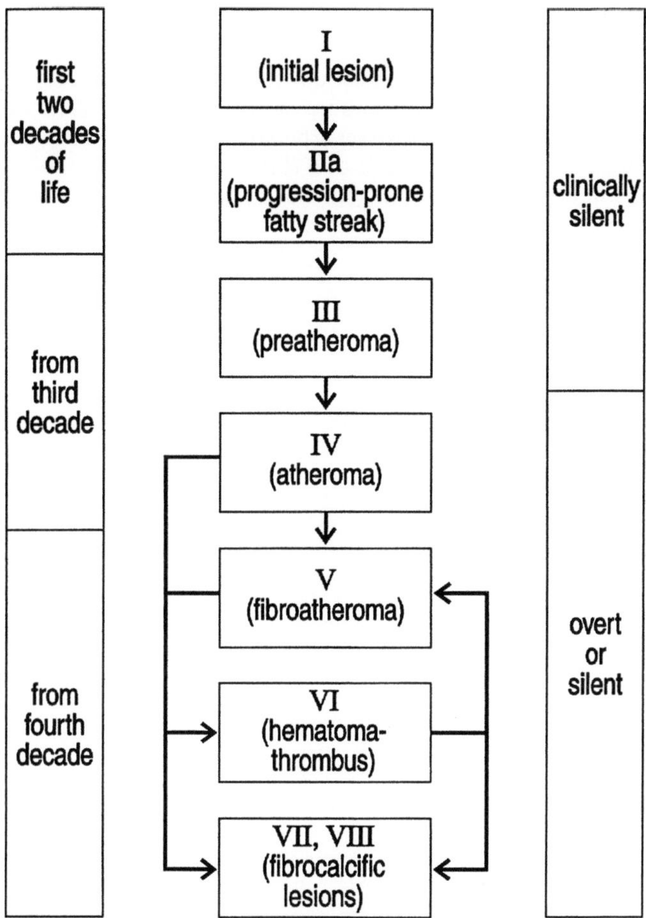

Fig. 3. The diagram indicates the sequence in the evolution of Type I to Type IV lesions and the various possible sequences in the progression to lesion types beyond Type IV.

Mineralization takes the form of calcium phosphate and apatite replacing the accumulated remnants of dead cells and extracellular lipid. Rather than causing further enlargement of lesions, mineralization may add permanence to arterial deformity.

Variable amounts of mineral are present in most advanced lesions. With refined microscopic methods, even the Type IV lesions of young adults may reveal small aggregates of crystalline calcium phosphate among the lipid particles of lipid cores and within the cytoplasm of smooth muscle cells trapped and injured within lipid cores (32). Since many, perhaps most, advanced lesions contain some mineral deposits from their onset, the Type VII classification is appropriate only when mineralization dominates the picture. However, it is understood that lesion components such as lipid deposits and increased fibrous tissue are generally also present.

Type VIII (fibrotic) lesion

Some atherosclerotic lesions may consist entirely or almost entirely of dense collagen. The lipidic component here is minimal or absent. The evolution of the fibrotic lesion is obscure. Lipid may have been resorbed or it may never have been present. A fibrotic lesion may be the consequence of a thrombotic extension of a lipidic lesion with the extension converted to collagen. Such lesions may obstruct the lumen of medium-sized arteries severely and may even be occlusive.

Acknowledgements: The observations and conclusions reported in this article are based on studies by the author that have been supported by the National Institutes of Health (grant HL-22739).

References

1. Barger AC, Beeuwkes R III, Lainey LL, Silverman KJ (1984) Hypothesis: Vasa vasorum and neovascularization of human coronary arteries. A possible role in the pathophysiology of atherosclerosis. N Engl J Med 310: 175–177
2. Beeuwkes R III, Barger AC, Silverman KJ, Lainey LL (1990) Cinemicrographic studies of the vasa vasorum of human coronary arteries. In: Glagov S, Newman WP, Schaffer SA (eds) Pathobiology of the human atherosclerotic plaque. Springer-Verlag, New York, pp 425–432
3. Bjorkerud S (1969) Reaction of the aortic wall of the rabbit after superficial, longitudinal, mechanical trauma. Virchows Arch A (Pathol Anat) 347: 197–210
4. Bjorkerud S, Bondjers G (1973) Arterial repair and atherosclerosis after mechanical injury: Tissue response after induction of a larger superficial transverse injury. Atherosclerosis 18: 235–255
5. Caro CG, Parker KH, Fish PJ, Lever MJ (1985) Blood flow near the arterial wall and arterial disease. Clin Hemor 5: 849–871
6. Constantinides P (1966) Plaque fissuring in human coronary thrombosis. J Atheroscl Res 6: 1–17
7. Constantinides P (1990) Plaque hemorrhages, their genesis and their role in supra-plaque thrombosis and atherogenesis. In: Glagov S, Newman WP, Schaffer SA (eds). Pathobiology of the human atherosclerotic plaque. Springer-Verlag, New York, pp 394–411
8. Cornhill JF, Herderick EE, Stary HC (1990) Topography of human aortic sudanophilic lesions. Monographs on Atherosclerosis, Volume 15, pp 13–19, S. Karger, Basel
9. Davies MJ, Thomas AC (1985) Plaque fissuring – the cause of acute myocardial infarction, sudden ischemic death and crescendo angina. Br Heart J 53: 363–373
10. Fuster V, Badimon L, Badimon JJ, Chesebro JH (1992) The pathogenesis of coronary artery disease and the acute coronary syndromes. N Engl J Med 326: 242–250; 310–318
11. Geer JC, Malcom GT (1965) Cholesterol ester fatty acid composition of human aorta fatty streaks and normal intima. Exp Mol Pathol 4: 500–507
12. Guyton JR, Klemp KF, Mims MP (1991) Altered ultrastructural morphology of self-aggregated low density lipoproteins: coalescence of lipid domains forming droplets and vesicles. J Lipid Res 32: 953–962

13. Katsuda S, Boyd HC, Fligner C, Ross R, Gown AM (1992) Human Atherosclerosis. III. Immunocytochemical analysis of the cell composition of lesions of young adults. Am J Pathol 140: 907–914

14. Katz SS, Shipley GG, Small DM (1976) Physical chemistry of the lipids of human atherosclerotic lesions. Demonstration of a lesion intermediate between fatty streaks and advanced plaques. J Clin Invest 58: 200–211

15. Mauer AM (1986) Risk factors in children and early atherosclerosis (letter to the editor). N Engl J Med 314: 1579

16. Mauer AM (1987) Atherosclerosis (letter to the editor). Pediatrics 79: 651–653

17. McGill HC (1974) The lesion. In: Schettler G, Weizel A (eds). Atherosclerosis III. Springer-Verlag, Berlin, pp. 27–38

18. McMillan GC (1985) Nature and definitions of atherosclerosis. In: Lee, KT (ed) Atherosclerosis. Ann NY Acad Sci 454: 1–4

19. Mitchell JRA, Schwartz CJ (1965) Arterial disease. F. A. Davis Co, Philadelphia

20. Møller L, Kristensen TS (1991) Plasma fibrinogen and ischemic heart disease risk factors. Arteriosclerosis and Thrombosis 11: 344–350

21. Munro JM, Van der Walt JD, Munro CS, Chalmers JAC, Cox EL (1987) An immunohistochemical analysis of human aortic fatty streaks. Hum Pathol 18: 375–380

22. Richardson M, Hatton MWC, Moore S (1988) Proteoglycan distribution in the intima and media of the aortas of young and aging rabbits: an ultrastructural study. Atherosclerosis 71: 243–256

23. Schmitz G, Müller G (1991) Structure and function of lamellar bodies, lipid-protein complexes involved in storage and secretion of cellular lipids. J Lipid Res 32: 1539–1570

24. Schwenke DC, Carew TE (1988) Quantification in vivo of increased LDL content and rate of LDL degradation in normal rabbit aorta occurring at sites susceptible to early atherosclerotic lesions. Circ Res 62: 699–710

25. Schwenke DC, Carew TE (1989) Initiation of atherosclerotic lesions in cholesterol-fed rabbits. I. Focal increases in arterial LDL concentration precede development of fatty streak lesions. Arteriosclerosis 9: 895–907

26. Small DM (1988) Progression and regression of atherosclerotic lesions. Arteriosclerosis 8: 103–129

27. Smith EB, Smith RH (1976) Early changes in aortic intima. In: Paoletti R, Gotto AM, Jr (eds). Atherosclerosis Reviews, vol. 1, Raven Press, New York, pp 119–136

28. Spring PM, Hoff HF (1989) LDL accumulation in the grossly normal human iliac bifurcation and common iliac arteries. Exp Mol Pathol 51: 179–185

29. Stary HC (1974) Proliferation of arterial cells in atherosclerosis. In: Wagner WD, Clarkson TC (eds). Arterial mesenchyme and arteriosclerosis. Plenum Press, New York, Adv Exp Med Biol 43: 59–81

30. Stary HC (1987) Macrophages, macrophage foam cells, and eccentric intimal thickening in the coronary arteries of young children. Atherosclerosis 64: 91–108

31. Stary HC (1989) Evolution and progression of atherosclerotic lesions in coronary arteries of children and young adults. Arteriosclerosis 9 (Suppl I): 19–32

32. Stary HC (1990) The sequence of cell and matrix changes in atherosclerotic lesions of coronary arteries in the first forty years of life. Eur Heart J 11 (Suppl E): 3–19

33. Stary HC, Blankenhorn DH, Chandler AB, Glagov S, Insull W, Rosenfeld ME, Richardson M, Schaffer SA, Schwartz CJ, Wagner WD, Wissler RW (1992) A definition of the intima of human arteries and of its atherosclerosis-prone regions. Circulation 85: 391–405. Arteriosclerosis and Thrombosis 12: 120–134

34. Steinberg D, Witztum JL (1990) Lipoproteins and atherogenesis. Current concepts. JAMA 264: 3047–3052

35. Stemerman MB, Ross R (1973) Experimental arteriosclerosis: 1. Fibrous plaque formation in primates: An electron microscopic study. J Exp Med 136: 769–789

36. Wight TN, Ross R (1975) Proteoglycans in primate arteries. I. Ultrastructural localization and distribution in the intima. J Cell Biol 67: 660–674

37. Wright HP (1968) Endothelial mitosis around aortic branches in normal guinea pigs. Nature 220: 78–79
38. Yarnell JWG, Baker IA, Sweetnam PM, Bainton D, O'Brien JR, Whitehead PJ, Elwood PC (1991) Fibrinogen, viscosity, and white blood cell count are major risk factors for ischemic heart disease. Circulation 83: 836–844
39. Zarins CK, Giddens DP, Bharadvaj BK, Sottiurai VS, Mabon RF, Glagov S (1983) Carotid bifurcation atherosclerosis. Quantitative correlation of plaque localization with flow velocity profiles and wall shear stress. Circ Res 53: 502–514

Author's address:
Herbert C. Stary, M.D.
LSU Medical Center
1901 Perdido Street
New Orleans, LA 70112
USA

Lipid and cellular constituents of unstable human aortic plaques

M. J. Davies[1], N. Woolf[2], P. Rowles[2], P. D. Richardson[3]

[1]British heart foundation cardiovascular pathology unit, St. George's Hospital
Medical School, Cranmer Terrace, London;
[2]Bland-Sutton Institute of Pathology, Middlesex Hospital Medical School, Riding
Housestreet, London;
[3]Brown University, Division of Engineering, Providence, Rhode Island, USA

Summary: Unstable plaques are undergoing thrombosis which, in most instances, is due to fissuring and rupture of the plaque cap. This process (deep intimal injury) is a complication of plaques with a lipid-rich core. The cap tear allows blood to enter the core from the lumen, leading initially to intraplaque thrombosis and, subsequently, in some cases intraluminal thrombosis.

Cap tears reflect the interplay between the force exerted on the tissue and its inherent mechanical strength.

Factors which elevate and concentrate circumferential wall stress on the cap during systole include an increasing proportion of the total plaque volume occupied by the lipid core, thinning of the cap and a loss of internal collagen struts within the core.

Factors which lead to an inherent reduction in the mechanical strength of cap tissue include a reduction in collagen and glycosaminoglycan concentrations, an increase in the number and density of macrophages, and a concomitant reduction in smooth muscle cells in the cap tissue.

It is therefore possible to define a vulnerable plaque as one in which the lipid core is disproportionately large, the cap thin, and in which monocytes preponderate over smooth muscle cells.

Key words: Plaque fissuring – thrombosis – unstable plaque – vulnerable plaque

The fact that most myocardial acute ischaemic episodes are precipitated by coronary thrombosis is now accepted (8, 9), and has been established by pathology, angiography (2, 5), and angioscopy (7, 12). In turn, these facts highlight the question of why an atherosclerotic plaque enters an unstable phase, which results in thrombosis. Pathological studies have shown that the majority of episodes of thrombosis which are large enough to result in luminal obstruction follow the tearing or cracking of the cap of a plaque. The consequence of plaque cap tearing is that the lipid core is exposed to blood in the arterial lumen (4, 5, 6). The lipid core is rich in tissue factor, lamellar lipid and exposed collagen; in brief, the interior of a lipid rich plaque is an extremely potent thrombogenic agent (19). Pathological studies have emphasised that the type of plaques which fissures has a necrotic lipid rich core (14, 18), and it is into this that a tear extends from the lumen transecting the fibromuscular tissue of the plaque cap. The fact that plaques which have undergone thrombosis usually have such a lipid core is now confirmed both by angioscopy (12) and by intravascular ultrasound in living patients (13, 15, 17). There is however little data, either on the size of the lipid pool as a proportion of total plaque volume which constitutes a high risk of fissuring, or on the quantitative changes in cell populations which predispose to fissuring.

Most patients with atherosclerosis have several, or even many, coronary, carotid and aortic plaques containing lipid, and there must be additional reasons for any individual plaque to rupture. In the coronary arteries plaques rupture, and heal, over a relatively short period of time. In contrast, the rather larger plaques of the carotid artery and aorta in man also undergo rupture but healing is prolonged, and it is by no means unusual to find at necropsy more than one plaque on which thrombosis is superimposed over an area of disruption in the plaque cap. The term ulcerated plaque is often used to indicate this chronicity in the carotid artery or aorta (1). At necropsy it is therefore possible in the aorta of the same individual to compare plaques which have become unstable and ruptured with those that have not and are still in an intact stable state.

We have carried out a study which provides quantitative data on the lipid and cellular components of intact compared with ruptured aortic plaques. Three groups of plaques were examined. Group A were intact plaques taken from the aortas of subjects without any ruptured (ulcerated) plaques. That is, they came from aortas in which all the plaques were in a stable phase. Group B plaques were intact, but came from aortas in which other plaques (Group C) were ruptured, and had invoked thrombosis. Thus plaques in Group C were unstable while presumably amongst group B were some plaques which might be becoming unstable. For every plaque the size of the extracellular pool of lipid was measured by quantitative microscopy. In a similar way the proportion of the cap tissue occupied by smooth muscle cells and macrophages was measured.

The volume of the plaque occupied by extracellular lipid showed a marked rise from A to B to C (Fig. 1). Consideration of the individual data points suggests that a cut-off point where the lipid core occupies more than 40 % of the total plaque volume

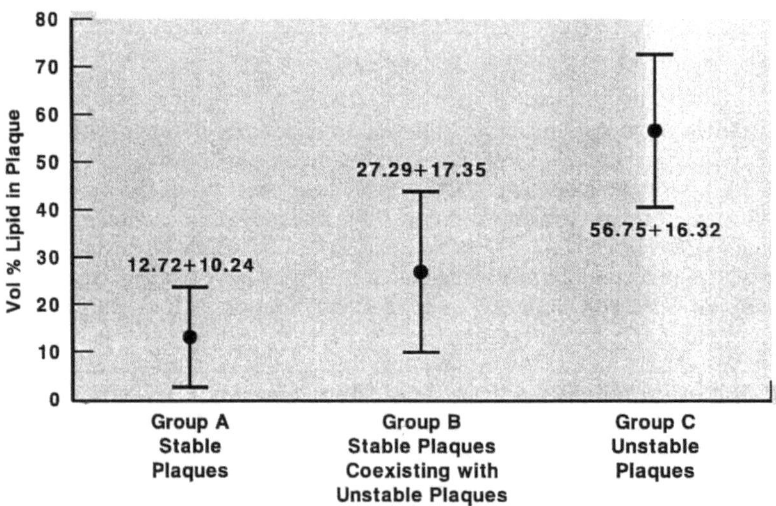

Fig. 1. Comparison of lipid content of plaque types. There is a steady increase from A to B to C with over 90 % of unstable plaques having a lipid component of over 40 % by volume.

exists; above this level plaque rupture is likely. Comparison of the cell composition of the cap tissue itself showed a decline in the volume occupied by smooth muscle cells (SMC), and an increase in the volume occupied by monocyte/macrophage derived foam cells (Mo) from A to B to C (Figs. 2, 3). In terms of the relative numbers of cells

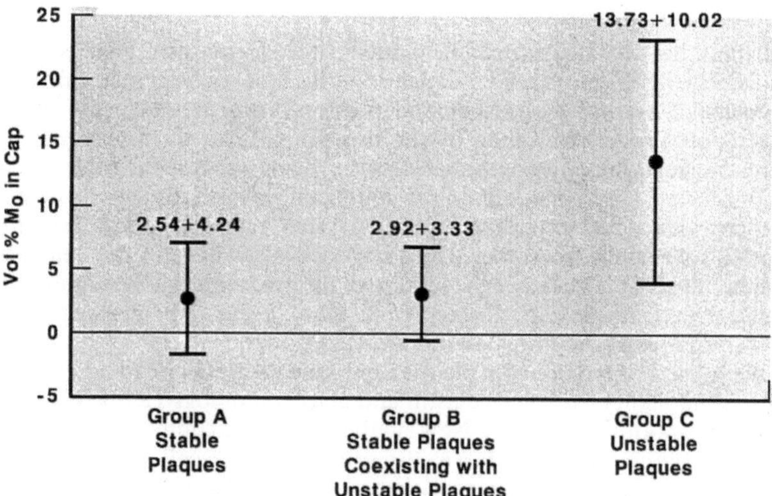

Fig. 2. Volume of cap tissue occupied by lipid filled macrophages in different plaque groups. Unstable plaques have a very much larger component of macrophages, measured by volume, in cap tissue.

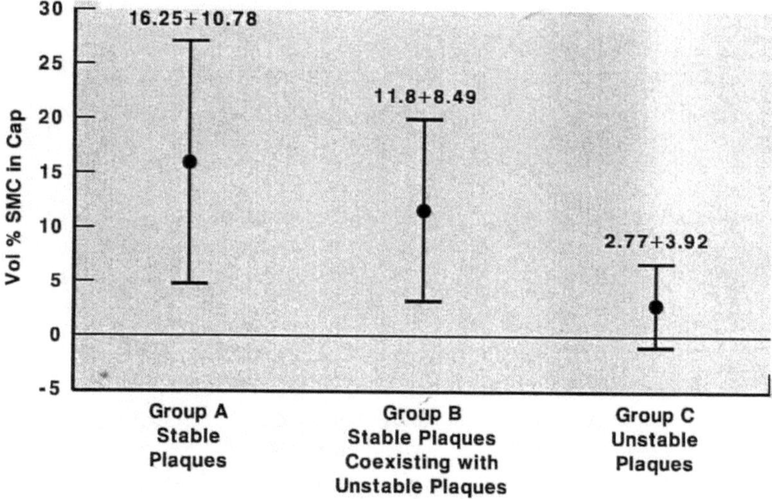

Fig. 3. Volume of cap tissue occupied by smooth muscle cells. Unstable plaque caps have a much reduced volume by smooth muscle cells.

35

the caps of intact plaques were dominated by smooth muscle cells (SMC), while the SMC/Mo ratio was reversed in ulcerated plaques to allow the preponderant cell to be the lipid containing foam cell (Fig. 4). Increasing proportions of the plaque volume occupied by lipid were associated with an increase in total monocyte foam cell numbers in the cap tissue.

It is therefore possible to categorise unstable plaques as being lipid-rich, monocyte/foam cell-rich, and smooth muscle cell depleted. It becomes an important question as to whether these changes can be reversed; limited data from animal models so far suggests that they can (3, 16).

These quantitative data on unstable plaques fit well with other studies concerning the pathogenesis of plaque rupture. A plaque with a lipid-rich necrotic core has a soft, easily deformable mass of cholesterol, and its esters, encapsulated in the intima. This mass is separated from the lumen by the innermost layer of the plaque, the cap. There is a single layer of endothelial cells covering the cap but the mechanical strength of the tissue is dependent on the connective tissue structure proteins (collagen, elastin) produced by smooth muscle cells. At each systolic pulse considerable circumferential wall stress, more than that imposed elsewhere in the vessel wall, is placed on the cap (14). The larger the lipid pool the greater is the circumferential wall stress concentrated on the cap. *Thus cap tissue is subject to an enhanced mechanical load which increases as the size of the lipid core increases.*

The mechanical strength of the plaque cap tissue itself is dependent on its connective tissue components, collagen, elastin and glycosoaminoglycans (GAGs); all are synthesised by smooth muscle cell. New synthesis of these connective tissue extracellular proteins components must fall as the smooth muscle cell population declines. Biochemical assays (Table 1) suggest that a decrease in collagen and GAGs are indeed a feature of ulcerated plaque caps. *Thus cap tissue may be mechanically weak*

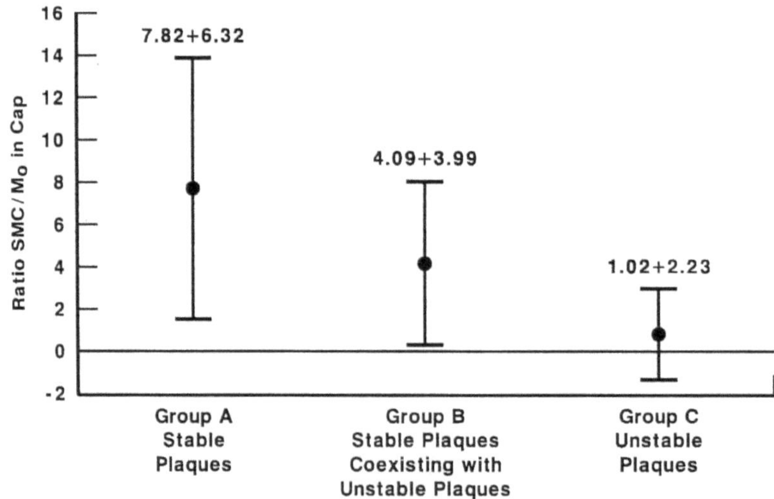

Fig. 4. Ratio of number density of smooth muscle cells and macrophages in cap tissue. Stable plaques are characterised by a considerable excess of smooth muscle cells. In unstable plaques the ratio of SMC/MO is unity or less.

Table 1. Biochemical analysis of pooled lesion-free intima, pooled non-ulcerated and pooled ulcerated plaque caps from seven individual patients. Results are mean ± SEM.

	Intima	Non-Ulcerated Cap	Ulcerated Cap
Total protein	59.9 ± 1.3	57.7 ± 2.2	54.8 ± 1.2
Collagen	41.4 ± 3.4	56.8 ± 1.4	35.4 ± 8.4*
Des + Ide	1.40 ± 0.26	1.17 ± 0.31	0.87 ± 0.27
Hexosamine	2.5 ± 0.1	2.4 ± 0.1	0.9 ± 0.5*
Total GAG	1.9 ± 0.2	1.9 ± 0.2	0.9 ± 0.2*
sGAG	1.6 ± 0.1	1.7 ± 0.1	0.9 ± 0.1*
C-4-S + C-6-S	1.0 ± 0.1	1.1 ± 0.1	0.6 ± 0.1*
Dermatan-S	0.5 ± 0.1	0.6 ± 0.02	0.2 ± 0.1*

All results expressed as % dry defatted weight except for Desmosine (Des) plus Isodesmosine (Ide) which are $\times 10^4$ nmol/g dry defatted weight. sGAG, sulphated glycosaminoglycan; C-4-S, chondroitin-4-sulphate; C-6-S, chondroitin-6-sulphate; Dermatan-S, dermatan sulphate. In paired t-tests between ulcerated and non-ulcerated plaque caps $p = < 0.05$*

if connective tissue protein content is reduced. In addition to a decline in the new production of connective tissue matrix, there may also be an active destruction of matrix proteins by lipid filled foam cells of monocyte/macrophage origin. Such lipid filled cells, which are, in effect, activated macrophages, have been shown to produce proteolytic enzymes within human plaques capable of connective tissue destruction (10). Plaque caps infiltrated by macrophages are mechanically weak (11). *Thus connective*

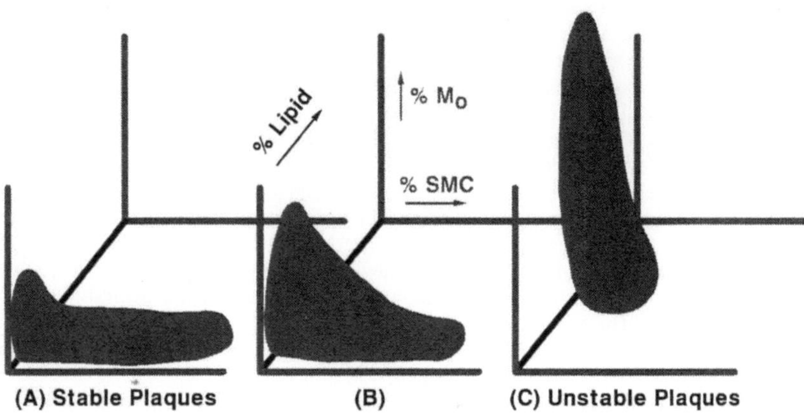

Fig. 5. Three-dimensional plots relating extracellular lipid, smooth muscle content of cap and macrophage content of cap. The dark areas include 95 % of all the plotted points. Stable plaques are characterised by a low lipid content (close to front of plot), a low macrophage and high smooth muscle content. In contrast unstable plaques have a high lipid component and a reverse in the relative content of smooth muscle cells and macrophages.

tissue proteins in cap tissue are reduced because of a combination of decreased synthesis and increased degradation.

Plaque instability is thus a reflection of increasing lipid deposition, increasing macophage numbers, and a decline in the smooth muscle cell population. The latter can be seen as a failure of the normal repair process within the intima, and a predominance of the destructive inflammatory processes. It remains to be seen whether therapeutic measures can reverse the imbalance of these processes. Smoth muscle proliferation, so often regarded as harmful as a cause of post-angioplasty stenosis in another context, can also be seen to be beneficial and essential for plaque stability.

References

1. Amarenco P, Duychaerts C, Tzourio C, Henn D, Bousser MG, Hauw JJ (1992) The prevalence of ulcerated plaques in the aortic arch in patients with stroke. N Engl J Med 326: 221–225
2. Ambrose J, Winters S, Arora R et al. (1985) Coronary angiograph morphology in acute myocardial infarction: link between the pathogenesis of unstable angina and myocardial infarction. JACC 6: 1233–1238
3. Blankenhorn D, Kramsch D (1989) Reversal of atherosis and sclerosis. The two components of atherosclerosis. Circulation 79: 1–7
4. Constantinides P (1966) Plaque fissures in human coronary thrombosis. J Atheroscl Res 6: 1–17
5. Davies M, Thomas A (1985) Plaque fissuring – the cause of acute myocardial infarction, sudden ischaemic death and crescendo angina. Br Heart J 53: 363–373
6. Falk E (1985) Unstable angina with fatal outcome: dynamic coronary thrombosis leading to infarction and/or sudden death. Circulation 71: 699–708
7. Forrester J, Litvak F, Grundfest W, Hickey A (1987) A perspective of coronary disease seen through the arteries of a living man. Circulation 75: 505–513
8. Fuster V, Badimon L, Badimon J, Chesebro J (1992) Mechanisms of Disease: The pathogenesis of coronary artery disease and the acute coronary syndromes I. N Engl J Med 326: 242–250
9. Fuster V, Badimon L, Badimon J, Chesebro J (1992) The pathogenesis of coronary artery disease and the acute coronary syndrome II. N Engl J Med 326: 310–318
10. Henney A, Wakeley P, Davies M et al. (1991) Localization of stromelysin gene expression in atherosclerotic plaques by in situ hybridization. Proc Natl Acad Sci USA 88: 8154–8158
11. Lendon C, Davies M, Born G, Richardson P (1991) Atherosclerotic plaque caps are locally weakened when macrophages density is increased. Atherosclerosis 87: 87–90
12. Mizuno K, Satomura K, Miyamoto A, Arakawa K, Shibuya T, Arai T, Kurita A, Nakamura H, Ambrose JA (1992) Angioscopic evaluation of coronary-artery thrombi in acute coronary syndromes. N Engl J Med 326: 287–291
13. Nissen S, Gurley J, Booth D et al. (1991) Differences in ultravascular plaque morphology in stable and unstable patients. Circulation 84: 436
14. Richardson P, Davies M, Born G (1989) Influence of plaque configuration and stress distribution on fissuring of coronary atherosclerotic plaques. Lancet II: 941–944
15. Rosenfeld M (1991) Oxidized LDL affects multiple atherogenic cellular responses. Circulation 83: 2137–2140
16. Small DM, Bond MG, Waugh D, Prack M, Sawyer JK (1984) Physiochemical and histological changes in the arterial wall of nonhuman primates during progression and regression of atherosclerosis. J Clin Invest 73: 1590–1605

17. Tobis J, Mallery J, Mahon D et al. (1991) Intravascular ultrasound imaging of human coronary arteries in vivo: analysis of tissue characterizations with comparison to in vitro histological specimens. Circulation 83: 913–926
18. Tracy R, Devaney K, Kissling G (1985) Characteristics of the plaque under a coronary thrombus. Virchows Arch Pathol Anat 405: 411–427
19. Wilcox J (1991) Analysis of local gene expression in human atherosclerotic plaques by in-situ hybridisation. Trends Cardiovasc Med 1: 17–24

Author's address:
Prof. M. J. Davies
BHF Cardiovascular Pathology Unit
St. George's Hospital Medical School
Cranmer Terrace, London SW17 ORE
UK

Immunological control mechanisms in plaque formation

G. K. Hansson

Gothenburg University, Department of Clinical Chemistry, Sahlgren's Hospital, Gothenburg, Sweden

Summary: Atherosclerosis is characterized by cholesterol accumulation, inflammation, and fibrous tissue formation. We have analyzed inflammatory components of atherosclerotic plaques and obtained evidence for T lymphocyte activation and cytokine secretion. A molecular genetical characterization of T cell clones obtained from atherosclerotic lesions revealed that the cells are heterogeneous with regard to antigen receptor gene organization. This indicates that they are derived from several progenitors and respond to different antigenic epitopes. The latter are not yet known, and it is also unclear to what extent the lymphocytic infiltrate in plaques represent a local immune response.

Vascular effects of cytokines produced by plaque macrophages and lymphocytes were studied in cell culture and animal experiments. It was found that the T cell cytokine, interferon-γ, inhibits cholesterol accumulation and foam cell formation by down-regulating the scavenger receptor on macrophages. It also inhibits smooth muscle proliferation in culture and the formation of arterial restenosis after angioplasty in experimental animals. Together, these studies emphasize the importance of vascular-immune interactions in the pathogenesis of atherosclerosis.

Key words: Cytokines – contractility – cell proliferation – inflammation – macrophages – T lymphocytes

Introduction

Atherosclerosis is a slowly progressive disease that probably has a multifactorial origin. The three major characteristics of the atherosclerotic plaque are the accumulation of cholesterol, the formation of a fibrous cap of smooth muscle cells and connective tissue fibers, and the infiltration of inflammatory cells. The importance of the inflammatory component is underlined by the fact that most of the cholesterol-laden foam cells are derived from monocyte-macrophages (15). It is also interesting that atherosclerosis-like lesions develop rapidly and aggressively in organ transplants (28).

Cholesterol deposits are predominantly found in the core of the plaque and is largely intracellular in macrophage-derived foam cells. Other macrophages, together with T lymphocytes, are mixed with smooth muscle cells in the fibrous cap. This histologic organization raises important pathogenetic questions. Does the inflammatory infiltrate represent a specific immune response carried by antigen-specific T lymphocytes? What is the relationship between inflammation, lipid accumulation and smooth muscle proliferation? None of these questions can be answered completely at present, but recent data provide important clues to some of the questions.

Cellular organization of the plaque

Cholesterol deposits are predominantly found in the core of the plaque and are largely intracellular in macrophage-derived foam cells (2, 9, 15, 18, 32, 34). Other macrophages together with T lymphocytes are mixed with smooth muscle cells in the fibrous cap.

The presence of blood-borne cells such as monocyte-derived macrophages and T lymphocytes in vascular tissue does not necessarily imply that an active immune or inflammatory response is occurring. Instead, the cells may simply be trapped in a forming plaque composed of loose connective tissue elements and thrombi. It has, however, been shown that a large proportion of plaque T lymphocytes express activation markers, i.e. cell surface proteins whose expression is restricted to activated cells (11, 31, 32). This indicates that these cells have the capacity to secrete lymphokines and suggests that an active immune response may take place in the plaque. Further support for this hypothesis was obtained by the detection of interferon-γ protein in plaque tissue (11). Similarly, the macrophages of the plaque express histocompatibility proteins such as HLA-DR in high frequency, which is compatible with an activation also of these cells (11, 14, 32). The presence of mRNA and protein for macrophage-derived cytokines such as interleukin-1, TNF and PDGF in plaque tissue (22, 35, 36) supports this notion, although it should be kept in mind that other cells such as endothelial and smooth muscle cells may produce the same types of cytokines as the macrophages (21, 34).

Cytokine secretion and paracrine networks

Inflammatory responses occur in paracrine networks. An initial immune response results in cytokine secretion that recruits surrounding mesenchymal and immunocompetent cells into the reaction. These latter cell types also start to produce cytokines and express cytokine receptors, and the inflammatory response is amplified. The fact that cytokine genes are expressed in human atherosclerotic plaques suggests that a proinflammatory cytokine network is operating. Several of the cytokines exert potent effects on lipid metabolism, smooth muscle proliferation, endothelial anticoagulant properties and other important functions of the vessel wall, and this makes it reasonable to propose that local cytokine release is involved in the pathogenesis of atherosclerosis.

Some of the most striking effects of inflammatory cytokines is to regulate lipid metabolism. The best known case is for TNF. This cytokine downregulates lipoprotein lipase production in adipose tissue, leading to a reduction of fatty acid utilization for energy metabolism and to an increased VLDL level in blood. This effect is observed in systemic infections and in malignant diseases, and may be the major cause of cancer cachexia (4, 5). Interferon-γ, one of the major secretory products of activated T lymphocytes, downregulates lipoprotein lipase production by macrophages in a similar way as TNF does for adipocytes (17). It is therefore possible that local release of interferon-γ and TNF in the atherosclerotic plaque may affect local lipolysis and lipoprotein conversion (16).

The cholesterol-filled foam cells are the most characteristic features of the atheroslerotic plaque. Most of them are derived from macrophages (2, 5, 9, 18, 34), which accumulate cholesterol largely by receptor-mediated uptake of modified lipo-proteins (3, 6, 29). The scavenger receptor that mediates this uptake has been cloned and sequenced (19, 23) and this permits detailed studies of its regulation on the molecular level.

Recent studies have shown that the scavenger receptor is a cytokine-regulated gene. Both interferon-γ (8) and TNF (33) downregulate its expression by reducing its mRNA level. This results in a reduced lipoprotein uptake and a reduced cholesterol accumulation and inhibits the transformation of the macrophage into a foam cell (8). Instead, other functions of the macrophage are activated, particularly those related to its role in the immune system. For example, expression of histocompatibility genes is upregulated and oxygen radical production is enhanced (1). In this way, the macro-phage increases its antigen-presenting capacity and participates in local immune responses. It now appears that this occurs at the expense of a loss in lipid metabolic activity. From a cell biologic point of view, the cell may develop along either of two differentiation pathways. One is induced by cytokines such as interferon-γ and TNF and leads to macrophage priming, activation and antigen presentation. The other one is followed in the absence of "proinflammatory" cytokines and is characterized by avid lipoprotein uptake and lipase secretion. In the presence of oxidized lipopro-teins, this pathway would lead to the transformation of the macrophage into a foam cell.

It is now clear that foam cell formation is determined by at least two factors: the local cytokine secretion and the availability of modified lipoproteins that can be inter-nalized via the scavenger receptor(s). This emphasizes the importance of the local inflammatory response in the arterial wall, which can generate free oxygen radicals that modify lipoproteins as well as produce cytokines. The local inflammatory response may therefore be an interesting target for future anti-atherosclerotic drugs.

Cytokines regulate endothelial and smooth muscle proliferation

Cytokines are important growth regulators and the distinction between cytokines and growth factors is, in fact, rather vague. Thus, macrophages release both the pro-totypic growth factor, PDGF (platelet-derived growth factor) and the prototypic cytokine, interleukin-1. The latter may, however, also stimulate smooth muscle pro-liferation (20). This has recently been shown to be accomplished by an induction of autocrine PDGF production and PDGF-receptor expression (26). Cytokines and growth factors are therefore interconnected in complicated para- and autocrine net-works.

The major local growth promoting factors that have been identified in the atherosclerotic plaque include PDGF and interleukin-1 (21, 22, 27, 35, 36), which have been detected on the mRNA and/or protein level in plaque tissue. The best characterized growth inhibitors are transforming growth factor-β and interferon-γ, the latter of which has also been identified in plaque tissue by direct localization (11). Cell culture studies have shown that interferon-γ is a potent inhibitor of smooth muscle (10) and endothelial (7) cell proliferation and recent studies in experimental animals have demonstrated that interferon-γ inhibits smooth muscle proliferation after injury in vivo (12, 13). Therefore, interferon-γ might block restenosis after angioplasty and this could be a useful therapeutic strategy in clinical medicine (12).

The observation that cytokines produced by the immune system modulate tissue repair in injured blood vessels suggests that the immune system may control vascular repair processes. Strong support for this hypothesis was recently obtained when we observed that T-lymphocyte-deficient rats develop significantly larger arterial lesions after injury compared to normal, T-cell containing littermates (13).

Are there specific immune responses in atherosclerosis?

Immunohistochemical studies show that a substantial proportion of the macrophages and T lymphocytes in the atherosclerotic plaque are activated, and this suggests that they might react with specific antigens in the tissue. Such antigens have, however, not yet been identified and it is now important to determine whether this type of immune responses take place in atherogenesis.

Circumstantial evidence suggest that specific immune responses may be elicited against oxidation epitopes on lipoproteins. First, antibodies to such epitopes have been identified in man, but they can be detected in healthy individuals as well as in atherosclerotic patients (24). Second, prominent inflammatory infiltrates containing large numbers of B and T lymphocytes, macrophages and plasma cells develop around ceroid, i.e. local accumulations of oxidized lipids in the arterial adventitia (25). Again, it is unclear whether they are of general importance in atherosclerosis since they cannot be observed in all patients. It will now be important to analyze immune responses to oxidized lipids and lipoproteins in large epidemiologic studies to clarify whether they are associated with atherosclerosis. If that turns out to be the case, antibodies to such antigens might be useful diagnostic markers of the disease.

If immune responses to oxidized lipoproteins are important in atherosclerosis, one would expect them to be initiated in the atherosclerotic plaque where the concentration of oxidized lipoproteins is high. If that is the case, some of the T lymphocytes present in the plaque may recognize oxidation-induced epitopes on local lipoproteins. Recent molecular analyses of antigen receptor genes in plaque T lymphocytes have shown that these cells represent heterogeneous immune responses and not a few clones directed against a small number of epitopes (30). One would therefore expect a relatively large immune repertoire in the plaque, with T cells specific for a variety of antigenic epitopes, some of which may perhaps be generated by oxidation. Such specific T cell responses may, however, initiate an inflammatory reaction that would stimulate the immigration of other, "nonspecific" T cells as well as monocytes into the plaque. This would result in an inflammatory response with local cytokine secretion that can modulate local lipid metabolism and cell proliferation. If would, however, be expected to be rather complex from an immunologic point of view. It may therefore take some time before the interesting problem regarding the molecular characteristics of the inflammatory response in atherogenesis is solved.

44

References

1. Adams DO, Hamilton TA (1984) The cell biology of macrophage activation. Annu Rev Immunol 2: 283–318
2. Aqel NM, Ball RY, Waldmann H, Mitchinson MJ (1984) Monocytic origin of foam cells in human atherosclerotic plaques. Atherosclerosis 53: 265–271
3. Brown MS, Goldstein JL (1983) Lipoprotein metabolism in the macrophage: implications for cholesterol deposition in atherosclerosis. Annu Rev Biochem 52: 223–261
4. Cerami A, Beutler B (1988) The role of cachectin/TNF in endotoxic shock and cachexia. Immunol Today 9: 28–31
5. Cornelius P, Enerbäck S, Bjursell G, Olivecrona T, Pekala PH (1986) Regulation of lipoprotein lipase mRNA content in 3T3-L1 cells by tumor necrosis factor. Biochem J 249: 765–769
6. Fogelman AM, Schechter I, Seager J, Hokum M, Child JS, Edwards PE (1980) Malondialdehyde alteration of low density lipoprotein leads to cholesteryl ester accumulation in human monocyte-macrophages. Proc Natl Acad Sci USA 77: 2214–2218
7. Friesel R, Komoriya A, Maciag T (1987) Inhibition of endothelial cell proliferation by γ-interferon. J Cell Biol 104: 689–696
8. Geng Y-J, Hansson GK (1992) Interferon-γ inhibits scavenger receptor expression and foam cell formation in human monocyte-derived macrophages. J Clin Invest 89: 1322–1330
9. Gown AM, Tsukada T, Ross R (1986) Human atherosclerosis. II. Immunocytochemical analysis of the cellular composition of human atherosclerotic lesions. Am J Pathol 125: 191–207
10. Hansson GK, Jonasson L, Holm J, Clowes MM, Clowes AW (1988) γ-interferon regulates vascular smooth muscle proliferation and Ia expression in vivo and in vitro. Circ Res, 63: 712–719
11. Hansson GK, Holm J, Jonasson L (1989) Detection of activated T lymphocytes in the human atherosclerotic plaque. Am J Pathol 135: 169–175
12. Hansson GK, Holm J (1991) Interferon-γ inhibits arterial stenosis after injury. Circulation 84: 1266–1272
13. Hansson GK, Holm J, Holm S, Hedrich H, Fotev Z, Fingerle J (1991) T lymphocytes inhibit the vascular response to injury. Proc Natl Acad Sci USA 88: 10530–10534
14. Jonasson L, Holm J, Skalli O, Gabbiani G, Hansson GK (1985) Expression of class II transplantation antigens on vascular smooth muscle cells in human atherosclerosis. J Clin Invest 76: 125–131
15. Jonasson L, Holm J, Skalli O, Bondjers G, Hansson GK (1986) Regional accumulations of T cells, macrophages, and smooth muscle cells in the human atherosclerotic plaque. Arteriosclerosis 6: 131–138
16. Jonasson L, Bondjers G, Hansson GK (1987) Lipoprotein lipase in atherosclerosis: its presence in smooth muscle cells and absence from macrophages. J Lipid Res 28: 437–445
17. Jonasson L, Hansson GK, Bondjers G, Noe L, Etienne J (1990) Interferon-γ inhibits lipoprotein lipase in human monocyte-derived macrophages. Biochim Biophys Acta 1053: 43–48
18. Klurfeld DM (1985) Identification of foam cells in human atherosclerotic lesions as macrophages using monoclonal antibodies. Arch Pathol Lab Med 109: 445–449
19. Kodama T, Freeman M, Rohrer L, Zabrecky J, Matsudaira P, Krieger M (1990) Type I macrophage scavenger receptor contains α-helical and collagen-like coiled coils. Nature 343: 531–535
20. Libby P, Warner SJC, Friedman GB (1988) Interleukin-1: a mitogen for human vascular smooth muscle cells that induces the release of growth-inhibitory prostanoids. J Clin Invest 88: 489–495
21. Libby P, Warner SJC, Salomon RN, Birinyi LK (1988) Production of platelet-derived growth factor-like mitogen by smooth-muscle cells from human atheromata. N Engl J Med 318: 1493–1496
22. Libby P, Hansson GK (1991) Biology of disease. Involvement of the immune system in human atherogenesis: Current knowledge and unanswered questions. Lab Invest 64: 5–15

23. Matsukomoto A, Naito M, Itakura H et al. (1990) Human macrophage scavenger receptors: primary structure, expression and localization in atherosclerotic lesions. Proc Natl Acad Sci USA 87: 9133–9137
24. Palinski W, Rosenfeld ME, Ylä-Herttuala S, Gurtner GC, Socher SA, Butler SW, Parthasarathy S, Carew TE, Steinberg D, Witztum JL (1989) Low density lipoprotein undergoes oxidative modification in vivo. Proc Natl Acad Sci USA 86: 1372–1376
25. Parums DV, Chadwick DR, Mitchinson MJ (1986) The localization of immunoglobulin in chronic periaortitis. Atherosclerosis 61: 117–123
26. Raines EW, Dower SK, Ross R (1989) Interleukin-1 mitogenic activity for fibroblasts and smooth muscle cells is due to PDGF-AA. Science 243: 393–396
27. Ross R, Masuda J, Raines E et al. (1990) Localization of PDGF-B protein in macrophages in all phases of atherogenesis. Science 248: 1009–1012
28. Salomon RN, Hughes CCW, Schoen FJ, Payne DD, Pober JS, Libby P (1991) Human coronary transplantation-associated atherosclerosis. Am J Pathol 138: 791–798
29. Steinberg D, Parthasarathy S, Carew TE, Khoo JC, Witztum JL (1989) Beyond cholesterol. Modifications of low-density lipoprotein that increase its atherogenicity. N Engl J Med 320: 915–924
30. Stemme S, Rymo L, Hansson GK (1991) Polyclonal origin of T lymphocytes in human atherosclerotic plaques. Lab Invest 65: 654–660
31. Stemme S, Holm J, Hansson GK (1992) T lymphocytes in human atherosclerotic plaques are memory cells expressing CD45RO and the integrin VLA-1. Arterioscl Thromb 12: 206–211
32. van der Wal AC (1989) Atherosclerotic lesions in human. In situ immunophenotypic analysis suggesting an immune mediated response. Lab Invest 61: 166–172
33. Van Lenten BJ, Fogelman AM (1992) Lipopolysaccharide-induced inhibition of scavenger receptor expression in human monocyte-macrophages is mediated through tumor necrosis factor-a. J Immunol 148: 112–116
34. Vedeler CA, Nyland H, Matre R (1984) In situ charcterization of the foam cells in early human atherosclerotic lesions. Acta Pathol Microbiol Immunol Scand 92C: 133–137
35. Wang AM et al. (1989) Quantitation of mRNA by the polymerase chain reaction. Proc Natl Acad Sci USA 86: 9717–9721
36. Wilcox JN, Smith KM, Williams LT, Schwartz SM, Gordon D (1988) Platelet-derived growth factor mRNA detection in human atherosclerotic plaques by in situ hybridization. J Clin Invest 82: 1134–1139

Author's address:
Dr. G. K. Hansson
Gothenburg University
Department of Clinical Chemistry
Sahlgren's Hospital
S-41345 Gothenburg
Schweden

Proliferation versus atrophy – the ambivalent role of smooth muscle cells in human atherosclerosis

H.-E. Schaefer

Department of Pathology, University of Freiburg, FRG

Summary: Most of the current concepts on morphogenesis of atherosclerosis attribute the development of atherosclerotic lesions to the combined effects of two main cellular events: 1) activation of macrophages leading to lipoprotein phagocytosis by scavenger cells, and 2) proliferation of smooth muscle cells (SMC). SMC-like cells producing collagenous fibers and extracellular matrix are particularly involved in the formation of the so-called fibrous caps surrounding the core of an atheroma composed of foam cells and fatty debris. The fiber-forming SMC, in general, are said to result from a proliferation of media SMC which once have moved into the intima.

This view of origin of the fiber-forming SMC and the alleged proliferation of media SMC is mainly derived from experimental assays exposing the vessel wall to various kinds of physical or chemical injuries. It is the purpose of this paper to demonstrate that the results of those more or less ephemeral experiments differ from findings obtained from a combined histochemical and morphometric analysis of SMC in the aortic media in spontaneous human arteriosclerosis.

Instead of any proliferation, a significant atrophy of SMC occurs in the media with advancing age and progress of atherosclerosis. To some extent, this decrease in numerical and volumetric density of SMC is accompanied with intra- and extracellular calcification. It seems likely that the loss of contractile capacity of the media resulting from wasting of SMC, does slow down the stream of the interstitial fluid in the arterial wall. This stagnation must increase the life span of LDL moving through the interstitial space. The chemical alteration ensuing from aging of LDL mediates its binding to the scavenger receptors and uptake by macrophages.

So far, muscular atrophy of the media forms an atherogenic factor of its own, leading to final results similar to those as known from conditions of intravascular aging of LDL in hyperlipoproteinaemia. The augmentation of SMC-like cells in the intima is hardly to be derived from the atrophic media, but rather seems to be due to local proliferation of cells which, in the normal state, do occur in small numbers in the subendothelial space. These so-called myointimal or Langhans-cells share with SMC their content of α-actin, but they differ by their stellate configuration from the bipolar shape SMC of the media.

Key words: Leiomuscular atrophy – impairment of intramural fluid transport – myointimal Langhans cells

Introduction

Most of the current concepts on morphogenesis of atherosclerosis attribute the development of atherosclerotic lesions to the combined effects of two main cellular reactions:

1) receptor-mediated endocytosis of aged, hence chemically modified lipoproteins (4, 5, 12, 18, 24, 57) by intimal macrophages (34, 40, 42, 46, 48, 51) and
2) proliferation of fiber-forming smooth muscle cells (16, 19, 23, 42, 44, 59).

While there is little doubt that mesenchymal cells producing extracellular matrix (52, 53), collagenous (32, 33) and elastic (13) fibres (53) are particularly involved in the formation of the so-called fibrous cap surrounding the centrally placed lipid deposits of an atheroma, the origin of these cells remains a matter of debate. Based on the view of Wissler (59) and other authorities (for review [44]), these fiber-forming cells have been derived from proliferating smooth muscle cells migrating from the media to the intima. This view of origin seems to be confirmed by experimental assays exposing the vessel wall to various kinds of physical or chemical stress provoking a more or less pronounced cellular proliferation that eventually leads to arteriosclerosis-like lesions.

It is the purpose of this communication to demonstrate that the conclusions drawn from ephimeral animal experiments which, due to their prevailing short-term design, neglect the influences exerted by the complex process of aging, are hardly to be reconciled with the morphologic findings seen in spontaneous human atherosclerosis. Particular stress will be laid on the observation that with advanced age the smooth muscle cells of the media undergo severe atrophy rather than proliferation. As will be pointed out, muscular atrophy may exert adverse effects on the intramural transport of interstitial fluid, thus abetting local aging of lipoproteins and inducing further atherogenic phagocytotic events.

Morphogenesis of an atheroma by focal lipid accumulation

In an attempt to outline the two afore mentioned leading processes of spontaneous human atherosclerosis, at first the formation of an atheroma *sensu strictu* will be followed up.

In very early lesions (referred to as gelatinous elevations by Movat et al. (35), the intimal layer of the aorta shows a slight fibrotic thickening, and collagenous fibers become impregnated with diffusely distributed sudanophilic lipids that can de derived from a transendothelial transport of LDL (56) by transcytotic plasmalemmal vesicles (22, 39). LDL entering by this way into the interstitial space of the arterial wall develops a high affinity to collagen with advancing age-related chemical modification (20).

In a second step, aged lipoproteins primarily bound to collagen become targets for scavenger macrophages with increasing lipid load transforming themselves into foam cells. Small clusters of foam cells constitute the so-called fatty streaks according to the terminology of Movat et al. (35). In more advanced lesions, the clustered lipophagic foam cells are displaying few birefringent crystals, which indicate an incipient crystallization of cholesterol resulting from acid lipolysis of surplus LDL-derived esterified cholesterol.

The lipid-storing macrophages are rich in acid lipase and acid phosphatase, the enzyme-histochemical demonstration of which may be exploited to visualize the distribution and fate of these cells, provided that a cytochemical technique is applied which generates a colored reaction product insoluble in lipids. Progressive hydrolysis of LDL-derived cholesterol esters leads to an intracellular accumulation of large birefringent crystals of cholesterol monohydrate, which aggregate at their highest density in the oldest macrophages surrouding the center of an atheroma, while the younger foam cells at the periphery, which have not yet attained a critical cholesterol load, are mostly free of crystalline cholesterol. Finally, the overload of crystalline cholesterol obviously kills the macrophages. This process can be monitored by acid

48

phosphatase staining; macrophages stained red by this enzyme-histochemical reaction lose their outer cell borders and spill their lysosomal acid phosphatase into the necrotic center of the mature atheromatous lesion. The decay of macrophages frees the intracellular cholesterol crystals from their primary intralysosomal sites and contributes to the intrafocal concentration growing crystals of cholesterol monohydrate, phospholipids and other fatty residues which typically constitute the necrotic core of a full-blown atheromatous plaque. The appearance of the centrally placed cholesterol crystals depends on the mode of tissue preparation with slender profiles prevailing in sections, and flat rhomboid plates seen in smear preparations obtained from the gruel plaque. In short, an atheroma can be regarded as a graveyard of cholesterol collecting macrophages (47).

Proliferation of myointimal Langhans cells and fibrosis

With regard to local proliferation and fiber formation, a peculiar type of mesenchymal cell occurs in early fibrotic intimal lesions as well as in the fibrous caps of advanced atheromata. On vertical sections these cells appear as slender spindle-shaped cells containing alkaline phosphatase (46). On horizontal sections parallel to the endothelial surface, the same cells display conspicuous stellate profiles and form a reticular network that includes macrophages. These cells share with the smooth muscle cells of the media their content of alpha-actin (Fig. 1).

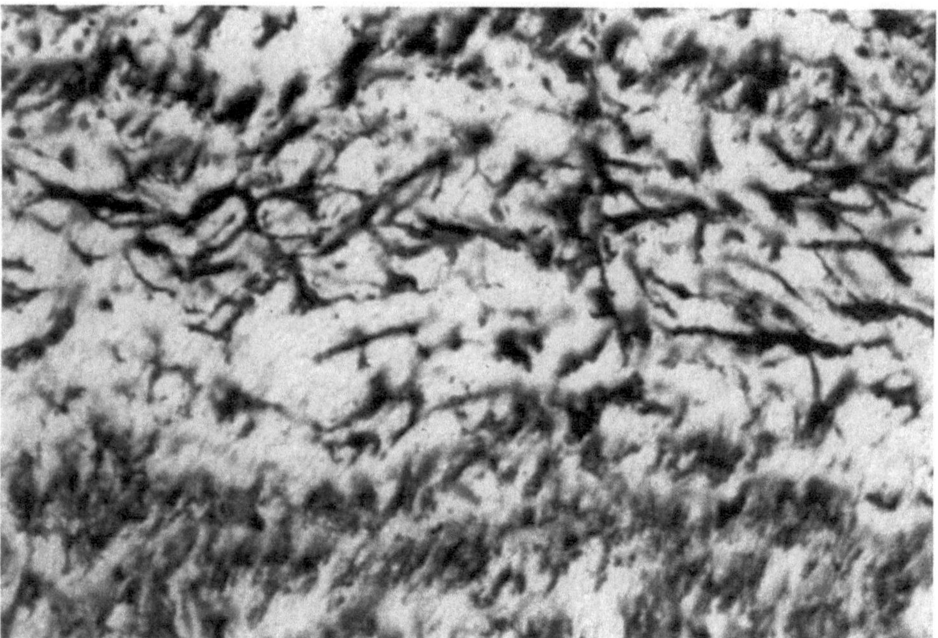

Fig. 1. Stellate myointimal Langhans-cells. Oblique incident section through the aortic intima of a 23-year-old man, immunostained for alpha-actin

As early as in 1866, Theodor Langhans (28), in full detail, described this peculiar cell type as occurring in normal as well as in arteriosclerotic vessels, and depicted them in his original drawings. Today, these smooth muscle-, as well as fibroblast-like stellate cells are often confused with the bipolar smooth muscles cells of the media and, likewise, the author's name happens to become erroneously mixed up with Paul Langerhans, the discoverer of the pancreatic islands and the intraepidermal reticular cell. To be historically correct, according to Thoma (58), the very first author who apostrophized to stellate intimal cells was Risse (41), who published his latin thesis in 1853.

Detailed descriptions of the myointimal Langhans-cells are also found in the more recent literature (8, 27, 37, 50). Electron-microscopic observations have revealed that the stellate Langhans cells are closely connected to the endothelium by tight junctions. Due to their subendothelial position, Langhans cells are candidates to receive various transendothelial signals and to react, e.g., stimulated by platelet-derived growth factor, with fiber production. With regard to their peculiar subendothelial position within the intima, the stellate smooth muscle-like Langhans cells may synonymously be referred to as so-called myointimal cells. Myointimal cells are prone to proliferation in atherosclerosis as can be readily demonstrated if sections obtained from various lesions of human atherosclerosis are immunostained for alpha-actin. Being aware of the existence of Langhans-cells and their inherent fiber-producing and proliferative capacity (13), we must not resort to the hypothesis that media myocytes proliferate and immigrate into the intima as a quasi *deus ex machina*.

Atrophy of smooth muscle cells in the tunica media

But the problem remains of what really happens to the smooth muscle cells of the media with advancing age and atherosclerosis. In order to answer this question, we have morphometrically analyzed the thoracic and abdominal aortae in 120 autopsied humans, covering the entire range from youth to older age (6, 49). To clearly identify smooth muscle cells, paraffin sections obtained from autopsy specimens were immunostained for smooth muscle alpha-actin by a modified (48) ABC-technique (21).

In normal aortae the brown stained smooth muscle cells are arranged in parallel layers with regularly interspersed elastic fibers (Fig. 2). However, with advancing age a progessive decrease of alpha-actin-positive cells occurs even outside atheromatous lesions. This muscular atrophy starts within the inner third of the media and may eventually lead to a complete fading of any alpha-actin-positive smooth muscle cells there. It is important to note that the morphometrical evaluation has been performed in areas outside atheromatous lesions in order to rule out any secondary influences exerted by an overlaying plaque (Fig. 3).

Leiomuscular atrophy of the media is a strictly age-dependent process which makes its appearance in early adulthood (Fig. 4). Atrophy may be defined as decrease, both of volumetric as well as of numerical density of alpha-actin-positive structures or smooth muscle cells, respectively. Emphasis must be laid on the finding that muscular wasting is not at all restricted to areas with overlaying atheromata and therefore does not result from emigration of smooth muscle cells from the media to the intima, as may be assumed following the aforementioned concept.

Leiomuscular atrophy of the media is more or less constantly combined with a fine granular calcification of the extracellular matrix, as demonstrated by a combina-

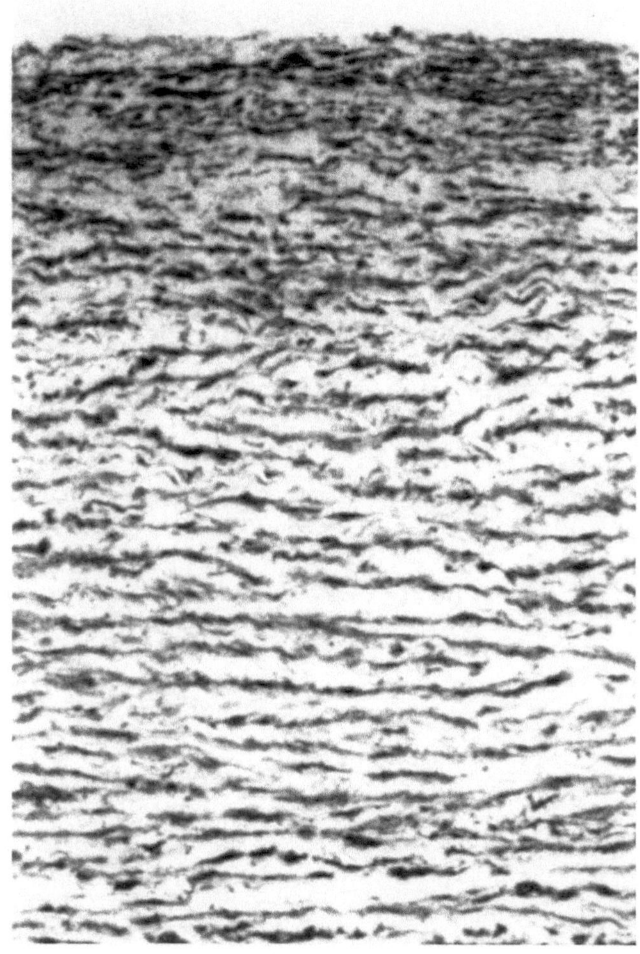

Fig. 2. Normal density of smooth muscle cells. Perpendicular section through intima and media of the aorta thoracalis of a 23-year-old man immunostained for alpha-actin

tion of immunostaining for alpha-actin with subsequent von Kòssa's reaction (26). In fact, the time-course monitored for leiomuscular atrophy of the media strictly parallels the age-dependent increase of calcium as measured by the group of Fleckenstein (14, 15) and other investigators (1, 7). This observation gives rise to the assumption that, somehow, cell damage is caused by primary calcium overload of smooth muscle cells (14, 15). However, even with the subtle technique of epifluorescent detection of individual silver grains resulting from the von Kòssa reaction, calcium deposits are seen only outside smooth muscle cells (Fig. 5). This finding is consonant with the view that matrix calcification primarily results from the advent of extracellular proteolipids, phospholipids (7, 43) and, perhaps, of calcium-binding proteins (29) of the gamma-carboxyglutaminic acid-containing type (30). Nevertheless, there may also occur a detrimental rise of the intracellular calcium concentration remaining below

51

Fig. 3. Leiomuscular atrophy in a 61-year-old man. Perpendicular section through intima and media of the aorta thoracalis (cf. Fig. 2)

the threshold of lightmicroscopical visibility. Staubesand and Seydewitz (55) have presented electronmicroscopical evidence in this respect.

Interference of leiomuscular atrophy with atherogenesis

It seems likely that the loss of contractile capacity of the media resulting from muscular atrophy will slow down the stream of interstitial fluid that, in the normal state, passes the arterial wall. This stagnation must increase the life span of LDL moving through the interstitial spaces. The chemical modifications known to ensue from aging of LDL mediates its binding to the scavenger receptors of macrophages, thus

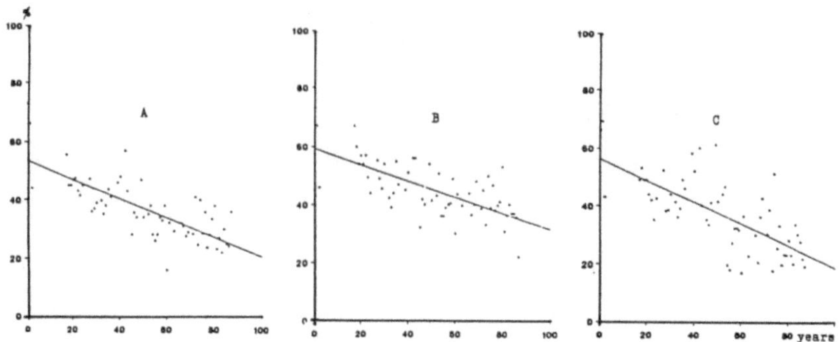

Fig. 4. Volumetric density of alpha-actin-positive smooth muscle cells in the inner (A), middle (B), and outer third (C) of the tunica media aortae thoracalis as morphometrically assayed by the point-counting method. There is a statistically significant decline which indicates an age-dependent leiomuscular atrophy. Similar results have been obtained from the abdominal aorta and for the numerical density of smooth muscle cells. Complete data are given elsewhere (6)

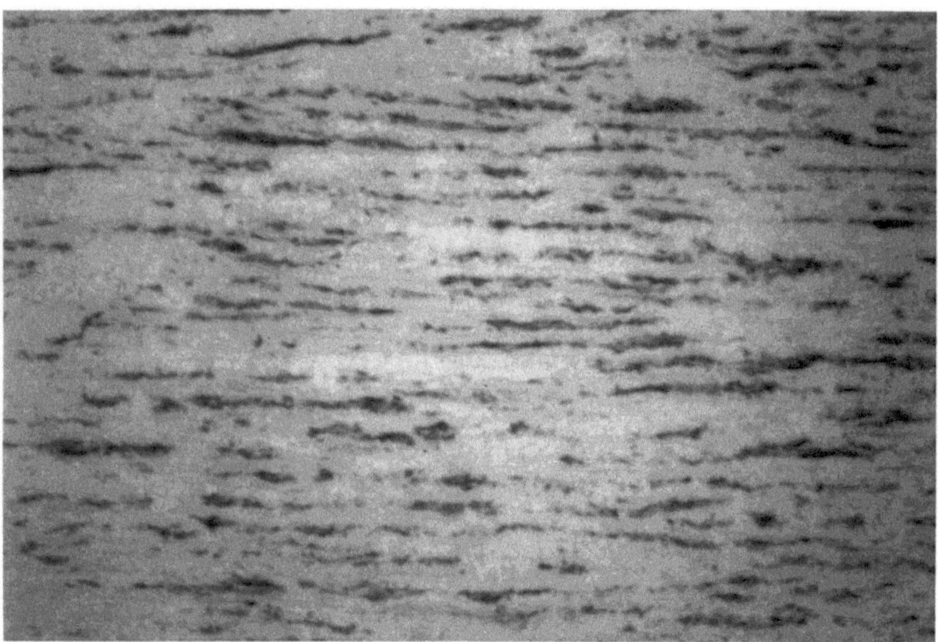

Fig. 5. Interstitial calcification and moderate atrophy of smooth muscle cells in the media aortae thoracalis of a 73-year-old woman. Smooth muscle cells are stained brown for alpha-actin, illumination by transmitting visible light; calcium deposits are stained green by the von Kòssa's reaction, epi-illumination by plane-polarized ultraviolet light

53

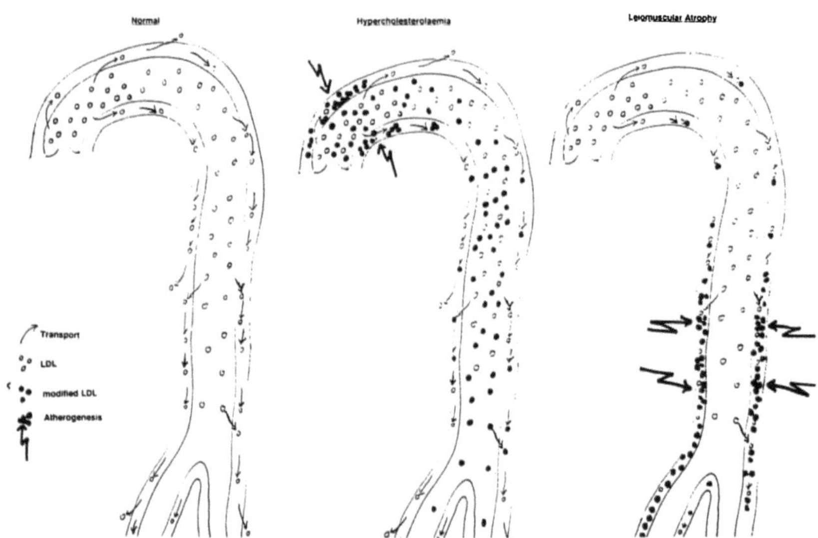

Fig. 6. Hypothetical diagram to explain the interaction of lipoprotein-dependent atherogenesis with leiomuscular atrophy. 1) In the normal state, LDL (and HDL) enters the vascular wall by transendocytosis and is quickly transported by an intact interstitial fluid stream without becoming bound to collagen or scavenger macrophages. 2) In hypercholesterolemia the intravascular life span of LDL becomes critically extended with an increased probability of intravascular chemical alteration; modified LDL is taken up by scavenger macrophages immediately after its entrance into the vessel wall; the resulting atheromas are located in the proximal aorta. 3) With increasing age-dependent leiomuscular atrophy and fibrosis, the interstitial stream of fluid is impeded to such an extent that chemical modification of LDL occurs on its way through the vessel wall; as a second step, chemically altered LDL will be intercepted by scavenger macrophages at places distant from its transendocytotic entrance; the resulting atheromas are mostly occupying the abdominal aorta

leading to receptor-mediated uptake, progressive cholesterol accumulation, and final decay of foam cells. In the normal state, LDL may pass the arterial wall without any chemical modification (Fig. 6). However, in severe hypercholesterolemia aging and chemical modification of LDL may occur inside the vascular space. In this setting, modified LDL is immediately taken up by macrophages when entering the vessle's wall. Under those conditions, atheromata will develop within the proximal aorta (Fig. 6). This particular type of proximal aortic atheromatosis is quite symptomatic for children and young adults suffering from familiar cholesterolemia (3, 4, 17, 38, 54) with its life span of intravascular LDL prolonged from 2.5 days at the normal, to 4.5 days in heterozygotes, and 6 days in homozygotes (4).

However, under the condition of age-dependent leiomuscular atrophy aging and chemical modification of LDL may occur inside the aortic wall where the stream of the interstitial fluid is impeded by the combined effects of leiomuscular atrophy, fibrosis, and calcification. Atheromata resulting from stagnant interstitial LDL are expected to develop somewhere in the course of intramural transportation, that means, prevailing in the more distal parts of the aorta and its branches (Fig. 6).

Atheromatosis of this distributional pattern is quite typical for the elderly (2, 10, 25, 45).

The basic assumption of a stream of interstital fluid floating LDL and other substances through the wall of arterial vessels seems definite (31). This stream is directed from the heart to the periphery ("a corde in peripheriam" [10]) due to the propulsive force of the pulse wave acting on the arterial wall like the rotating piston of a tubing pump. The effects of the pulse wave imparted to the interstitial fluid depends on its direction, frequency, and pressure. In order to convert the pulsatile blood pressure into a propulsive force moving the interstitial fluid, the intraluminal pressure must be counteracted by the tensile strength of the media which greatly depends on the contractile capacity of its smooth muscle cells. Of course, collagenous and elastic fibers may play a similar mechanical role; any increase of these constituents will impede the movement of fluids. So far, fibrosis mostly accompanying smooth muscular atrophy represents an additional factor that impairs the interstitial transport of LDL-containing fluids. Unfortunately, velocity and magnitude of this very stream are difficult to define. To the best of our knowledge, there are only limited data available that throw some light into this problem. Therefore, Okuneff (36) has made attempts to visualize the intransendothelial and intramural transport of fluid by intravital injection of stains, and Linzbach (31) has provided interesting trial calculations.

In conclusion, muscular atrophy of the arterial media is regarded as a hitherto widely ignored pathogenetic hinge in atherogenesis which contributes to the formation of atheromatous plaques and influences their topographic distribution.

References

1. Andersson HC, McGregor DH, Tanimura A (1986) Mechanisms of calcification in atherosclerosis. In: Glagov S, Newman WP, Schaffer SA (eds) Pathobiology of the human atherosclerotic plaque. Springer, New York, pp 235—249
2. Aschoff L (1925) Über die Atherosklerose. In: Vorträge über Pathologie, gehalten an den Universitäten und Akademien Japans im Jahre 1924 von L. Aschoff. Gustav Fischer-Verlag, Jena, pp 62—84
3. Assmann G (1982) Lipidstoffwechsel und Atherosklerose. Schattauer-Verlag, Stuttgart
4. Brown MS, Goldstein JL (1984) How LDL receptors influence cholesterol and atherosclerosis. Scientific American 251: 58—66
5. Brown MS, Goldstein JL (1990) Scavenging for receptors. Nature 343: 508—509
6. Bürger B (1993) Die altersabhängige Atrophie der glatten Muskelzellen in der Media der Aorta. Eine immunhistochemisch-morphometrische Untersuchung. Inauguraldissertation Freiburg
7. Bürger M (1939) Die chemischen Altersveränderungen an Gefäßen. Z Neurol Psychol 167: 273-280
8. Campbell G, Campbell J (1985) Smooth muscle phenotype changes in arterial wall homeostatis: implications for the pathogenesis of atherosclerosis. Exp Mol Pathol 42: 139—162
9. Doerr W (1960) Morphologische Untersuchungen zur Entstehung der Aortensklerose. DMW 85: 1401—1405
10. Doerr W (1978) Arteriosclerosis without end. Principles of pathogenesis and an attempt at a nosological classification. Virchows Arch A Path Anat 380: 91—106
11. Doerr W (1989) Über den Krankheitsbegriff — dargestellt am Beispiel der Arteriosklerose. Sitzungsbericht der Heidelberger Akademie der Wissenschaften, Mathematisch-naturwissenschaftliche Klasse, Jahrgang 1989, 2. Abhandlung, Springer-Verlag, Berlin
12. Dressel HA, Deigner HP, Frübis J, Strein K, Schettler G (1990) LDL-metabolism of the arterial wall — new implications for atherogenesis. Z Kardiol 79: Suppl 3, 9—16

13. Feigl W (1979) Reaktionsformen der glatten Muskelzelle der menschlichen Arterienwand
    — ihre Bedeutung für die Atherosklerose. Wiener Med Wschr, Suppl 32: 1—19
14. Fleckenstein A, Frey M, Zorn J, Fleckenstein-Grün G (1990) Calcium, a neglected key
    factor in hypertension and arteriosclerosis. Experimental vasoprotection with calcium
    antagonists or ACE inhibitors. In: Laragh JH, Brenner BM (eds) Hypertension:
    Pathophysiology, Diagnosis, and Management. Raven Press, New York, chapter 32
15. Fleckenstein A, Frey M, Thimm F, Fleckenstein-Grün G (1990) Excessive mural calcium
    overload — a predominant causal factor in the development of stenosing coronary plaques
    in humans. Cardiovascular Drugs and Therapy 4: 1005—1016
16. Gabbiani G (1986) Cytoskeletal characterization of smooth muscle cells of human and
    experimental atherosclerotic plaques. In: Glagov S, Newman WP, Schaffer SA (eds)
    Pathobiology of the human atherosclerotic plaque. Springer, New York, pp 63—68
17. Goldstein JL (1972) The cardiac manifestations of the homozygous and heterozygous forms
    of familial type II hyperbetalipoproteinemia. Birth Defect 8: 202
18. Goldstein JL, Brown MS (1977) The low-density lipoprotein pathway and its relationship to
    atherosclerosis. Ann Rev Biochem 46: 897—930
19. Gown AM, Tsukada T, Ross R (1986) Human atherosclerosis. II. Immunocytochemical
    analysis of the cellular composition of human atherosclerotic lesions. Am J Pathol 125:
    191—207
20. Hoover GA, McCornick S, Kalant N (1988) Interaction of native and cell-modified low
    density lipoprotein with collagen gel. Arteriosclerosis 8: 525—534
21. Hsu SM, Raine L, Fanger H (1981) Use of avidin-biotin-peroxidase complex (ABC) in im-
    munoperoxidase techniques: A comparison between ABC and unlabeled antibody (PAP)
    procedure. J Histochem Cytochem 29: 577—580
22. Hüttner I, Boutet M, More RH (1973) Studies on protein passage through arterial en-
    dothelium. I. Structural correlates of permeability in rat arterial endothelium. Lab Invest
    288: 672—677
23. Jonasson L, Hom J, Skalli O, Bondjers G, Hansson GK (1986) Regional accumulations of
    T cells, macrophages, and smooth muscle cells in the human atherosclerotic plaque. Ar-
    teriosclerosis 6: 131—138
24. Jürgens G, Hoff HF, Chisolm III GM, Esterbauer H (1987) Modification of human serum
    low density lipoprotein by oxidation — characterization and pathophysiological implica-
    tions. Chem Phys Lipids 45: 315—336
25. Koch W (1928) Provinzielle Ausbreitung und Charakter der Arteriosklerosen im röntgen-
    anatomischen Bilde. Verh Dtsch Ges Pathol 23: 478—487
26. von Kòssa J (1901) Ueber die im Organismus künstlich erzeugbaren Verkalkungen. Beitr
    Pathol Anat Allg Pathol 29: 163—202
27. Krushinsky AV, Orekov AN, Smirnov VN (1983) Stellate cells in the intima of human aorta.
    Application of alkaline dissociation method in the analysis of the vessel wall cellular con-
    tent. Acta Anat 117: 266—269
28. Langhans Th (1866) Beiträge zur normalen und pathologischen Anatomie der Arterien. Vir-
    chows Arch Pathol Anat 36: 187—226
29. Levy R, Lian JB, Gallop P (1979) Atherocalcin, a gamma-carboxylglutamic acid containing
    protein. Biochem Biophys Res Commun 91: 41—49
30. Lian JB, Hanschka PV, Gallop PM (1978) Properties and biosynthesis of a vitamin K-de-
    pendent calcium-binding protein in bone. Fed Proc 37: 2615—2620
31. Linzbach AJ (1958) Die Bedeutung der Gefäßwandfaktoren für die Entstehung der Ar-
    teriosklerose. Verh Dtsch Ges Pathol 41: 24—41
32. McCullagh KG, Ballian G (1975) Collagen charaterization and cell transformation in
    human atherosclerosis. Nature 258: 73—75
33. McCullagh KG, Duance VC, Bishop KA (1980) The distribution of collagen types I, III and
    V (AB) in normal and atherosclerotic human aorta. J Pathol 130: 45—55
34. Michinson MJ, Carpenter KLH, Ball RY (1986) The role of macrophages in human
    atherosclerosis. In: Glagow S, Newman WP, Schaffer SA (eds) Pathobiology of the human
    atherosclerotic plaque. Springer-Verlag, New York, pp 121—128

35. Movat HZ, Haust MD, More RH (1959) The morphologic elements in the early lesions of arteriosclerosis. Am J Pathol 35: 93–101

36. Okuneff N (1926) Über die vitale Farbstoffimbibition der Aortenwand. Virchows Arch Pathol Anat 259: 685–697

37. Orekov, Karpova II, Tertov VV, Rudchenko SA, Andreeva ER, Krushinsky AV, Smirnov VN (1984) Cellular composition of atherosclerotic and uninvolved human aortic subendothelial intima. Light-microscopic study of dissociated aortic cells. Am J Pathol 115: 17–24

38. Orth M, Volk P, Niederhoff H, Böhm N, Sander C (1973) Kardiovaskuläre Xanthomatose im Kindesalter bei primärer Hyperlipoproteinämie Typ II. Med Welt 24: 595–599

39. Palade GE, Burns RR (1968) Structural modulations of plasmalemmal vesicles. J Cell Biol 37: 633–649

40. Pitas RE, Innerarity TL, Maley RW (1983) Foam cells in explants of atherosclerosis rabbit aortas have receptors for β- very low density lipoproteins and modified low density lipoproteins. Arteriosclerosis 3: 2–12

41. Risse (1853) Observationes quaedam de arteriarum statu normali atque pathologico. Diss. inaug., Regiomont.

42. Roessner A, Herrera A, Höning HJ, Vollmer E, Zwaldo G, Schürmann R, Sorg C, Grundmann E (1987) Identification of macrophages and smooth muscle cells with monoclonal antibodies in the human atherosclerotic plaque. Virchows Arch Pathol Anat A 412: 169–174

43. Romeo R, Augstyn JM, Mandel G, Daoud A (1986) Characterization of nucleating proteolipids from calcified and non-calcified atherosclerotic lesions. In: Glagov S, Newman WP, Schaffer SA (eds) Pathobiology of the human atherosclerotic plaque, Springer, New York, pp 251–262

44. Ross R (1986) The pathogenesis of atherosclerosis: An update. New Engl J Med 314: 488–500

45. Rühl A (1929) Über die Gangarten der Arteriosklerose. Provinzielle Ausbreitung und Charakter mit besonderer Berücksichtigung des röntgenanatomischen Bildes. Veröff Kriegs-Konstitutionspathol 5: 1–78

46. Schaefer HE, Assmann G (1980) Bedeutung der Makrophagen für die Genese der Arteriosklerose. Münchener Med Wschr, Suppl 5, 122: 228–238

47. Schaefer HE (1981) The role of macrophages in atherosclerosis. In: Schmalzl F, Huhn D, Schaefer HE (eds) Disorders of the Monocyte Macrophage System. Pathophysiological and Clinical Aspects, Springer-Verlag, Berlin, Vol 27, pp 121–130

48. Schaefer HE (1984) Methoden zur histologischen, zytologischen und zytochemischen Diagnostik von Blut und Knochenmark. In: Remmele W (Hrsg) Pathologie, Springer-Verlag Berlin Bd. 1, pp 435–452

49. Schaefer HE, Bürger B (1992) Immunhistochemische Untersuchungen zur Frage einer altersbedingten Atrophie der glatten Muskelzellen in der Aorta. In: Heinle H, Schulte H, Schaefer HE (Hrsg) Arteriosklerotische Gefäßerkrankungen. 5. Tagung der Deutschen Gesellschaft für Arterioseforschung, Vieweg (Edition Dino; 8), Braunschweig, pp 181–188

50. Schönfelder M (1969) Orthologie und Pathologie der Langhans-Zellen der Aortenintima des Menschen. Path Microbiol 33: 129–145

51. Small DM (1980) Summary of concepts concerning the arterial wall and its atherosclerosis lesions. In: Gotto AM, Smith LC, Allen B (eds) Atherosclerosis V., Springer-Verlag, Berlin, pp 520–524

52. Smith EB (1965) The influence of age and atherosclerosis on the chemistry of aortic intima. 2. Collagen and mucopolysaccharides. J Atheroscler Res 5: 241–248

53. Smith P, Heath D (1980) The ultrastructure of age-associated intimal fibrosis in pulmonary blood vessels. J Pathol 130: 247–253

54. Stanley P, Chartaud C, Davignon A (1965) Acquired aortic stenosis in a twelve-year-old girl with xanthomatosis. N Engl J Med 273: 1378–1381

55. Staubesand J, Seydewitz V (1982) Elektronenmikroskopische Untersuchungen an Koronararterien des Hundes nach Anwendung kardioplegischer und myokardprotektiver Lösungen. In: Just H, Tschirkov A, Schlosser V (Hrsg) Kalziumantagonisten zur Kardioplegie und Myokardprotektion in der offenen Herzchirurgie. Int. Symp., Breisach 1981. Thieme-Verlag, Stuttgart, pp 165−172
56. Stein O, Stein Y, Eisenberg S (1973) A radioautographic study of the transport of ^{125}J-labeled serum lipoproteins in rat aorta. Z Zellforsch 138: 223−237
57. Steinbrecher UP, Lougheed M, Kwan W-Ch, Dirks M (1989) Recognition of oxidized low density lipoprotein by the scavenger receptor of macrophages results from derivatization of apolipoprotein B by products of fatty acid peroxidation. J Biol Chem 264: 15216−15223
58. Thoma R (1883) Über die Abhängigkeit der Bindegewebsneubildung in der Arterienintima von den mechanischen Bedingungen des Blutumlaufes. Erste Mittheilung. Die Rückwirkung des Verschlußes der Nabelarterien und des arteriösen Ganges auf die Struktur der Aortenwand. Virchows Arch Path Anat Physiol Klin Med 93: 443−506
59. Wissler RW (1980) The artery wall and the pathogenesis of progressive atherosclerosis. In: Gotto AM, Smith LC, Allen B (eds) Atherosclerosis V. Springer, New York, pp 407−414

Author's address:
Prof. Dr. med. H.-E. Schaefer
Ludwig Aschoff-Haus
Pathologisches Institut der Universität Freiburg
Albertstr. 19
Postfach 214
D-79002 Freiburg
FRG

Extracellular matrix degrading metalloproteinases in the pathogenesis of arteriosclerosis

A. C. Newby, K. M. Southgate, M. Davies*)

Department of Cardiology and *Institute of Nephrology,
University of Wales College of Medicine, Heath Park, Cardiff, UK

Summary: We review the importance of extracellular matrix remodelling to the processes of vascular smooth muscle cell migration and proliferation that contribute to morphogenesis of the atherosclerotic plaque. In particular, the role of the matrix degrading metalloproteinase (MMP) family is discussed. This family of neutral, Zn^{2+}-requiring enzymes are capable, in principle, of degrading all matrix proteins. Their activity is tightly controlled, however, at the level of snythesis of the inactive zymogens, activation by limited proteolysis and binding to endogenous inhibitor proteins (TIMPs). Direct evidence is presented for the involvement of MMPs in proliferation and outgrowth of vascular smooth muscle cells from explants of rabbit aorta in vitro. This was obtained using two structurally-unrelated inhibitors of MMPs, Ro 31-4724 and Ro 31-7467, both of which inhibited proliferation of cells in a concentration-dependent manner, Ro 31-4724 also inhibited outgrowth. Rabbit aortic smooth muscle cells were further shown to release MMPs, namely a 95 and a 72 kDa gelatinases that were inhibited by Ro 31-4724 and Ro 31-7467. The evidence suggests that degradation of basement membrane by gelatinase is required for proliferation and outgrowth of these cells. The implications of these findings for the pathogenesis and treatment of atherosclerosis are also discussed.

Key words: Vascular smooth muscle – proliferation – migration

Introduction

Current theories emphasise the importance of growth factors and the proliferation of vascular smooth muscle cells (VSMC) in atheroma formation (25). Migration of VSMC from the media is also an important contributor to intimal thickening during vascular repair (3, 7), and hence, by implication, diffuse intimal thickening and fibrous plaque formation. In animal models of cholesterol feeding, monocyte infiltration and fatty streak formation clearly preceed the recruitment of VSMC into the plaque (6, 16). In adult humans, however, the sequence of these events may be reversed. Sites prone to atheroma formation are also preferred locations for diffuse intimal thickening, composed of VSMC without abundant macrophage-derived foam cells. Indeed, in postmortem specimens, more than half of the haemodynamically significant stenoses in human coronary arteries are of this type (9). In the aorta,

---

* The original work described herein was supported by grants from the Medical Research Council and the British Heart Foundation.

intimal thickening may be virtually uniform, whereas atheroma is localised. Moreover, there appear to be only approximately twice as many intimal VSMC per unit area in atheromatous compared to non-atheromatous areas (22). These data suggest that atheroma may form on areas of pre-existing intimal thickening and that migration of VSMC rather than proliferation may be the predominant event underlying morphogenesis of the plaque. These alternative routes to atheroma formation are depicted in Fig. 1.

Extracellular matrix remodelling is necessary for migration of VSMC and there is also circumstancial evidence for its direct involvement in regulating proliferation. For example, it has long been recognised that components of the normal vessel wall, particularly laminin and the heparan sulphate glycosaminoglycans, act as inhibitors of modulation of VSMC into proliferative phenotypes and proliferation itself (2, 10, 31). If so, matrix degrading enzymes may be able to relieve this inhibition. More recently, it has been shown that certain growth factors, including bFGF (26), TGFß (18) and thrombomodulin (15), bind to matrix components and so may be released in active form by matrix degrading enzymes.

During migration the VSMC would encounter a series of distinct extracellular matrices (Fig. 2) (1, 19, 33). Firstly, each cell is surrounded by a basement membrane, comprised of Type IV collagen, laminin and heparan sulphate proteoglycans. Next to

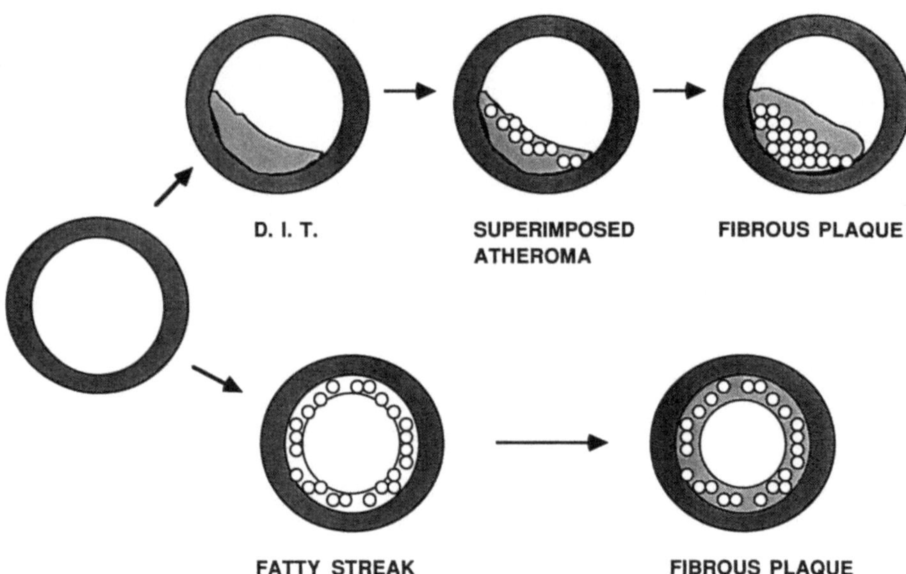

Fig. 1. Atheroma formation. Two hypothetical sequences of events that may lead to the development of atheroma. In the upper part of the figure progression from diffuse intimal thickening (D.I.T.) to atheroma may result from the infiltration of macrophages (O) with only a modest increase in vascular smooth muscle cell numbers. The plaque morphology is changed largely by migration of cells. This may reflect the pathology in man. In the lower part, macrophages encourage the migration and proliferation of smooth muscle cells from the media. This probably accounts for lesion formation in cholesterol-fed animals.

60

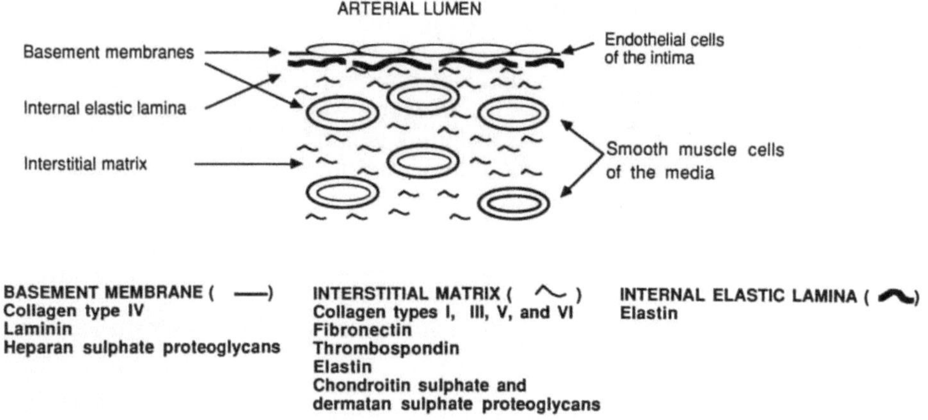

Fig. 2. Distribution of the types of extracellula matrix within the arterial wall.

be met would be the collagen fibrils (Types I, III, V and VI), fibronectin (which promotes phenotypic change) and proteoglycans of the interstitial matrix of the tunica media. The proteoglycans here are predominantly of the chondroitin sulphate and dermatan sulphate (decorin and biglycan) types. In order to enter the intima, cells must traverse the internal elastic lamina, which is rich in elastin. Although fenestrated, evidence from balloon catheter-induced injury to normal arteries suggests that the internal elastic lamina (IEL) is, nevertheless, a significant barrier. Specifically, Reidy and Silver (24) reported in a model of gentle injury that intimal proliferation occurred only at sites where the IEL was breached. A recent quantitative study in a pig model of angioplasty confirmed that intimal thickening was greater and more prolonged where the IEL was torn (8). Once in the intima, VSMC would then encounter the sub-endothelial basement membrane. Finally the matrix between the endothelial cells must represent a unique microenvironment for the VSMC because in vivo, the cells do not appear to migrate through or over the endothelium.

Matrix-degrading metalloproteinases (MMPs)

Among the different families of proteinases, the MMPs appear to play the pivotal role in the modulation of the extracellular matrix (5, 20, 34). At present, three subgroups have been recognised, namely interstitial collagenases, Type IV collagenases (gelatinases) and stromelysins (Table 1). These enzymes share several common features: I) they are optimally active at neutral pH, II) they have an absolute requirement for Zn^{2+} and, consequently, metal chelating agents such as EDTA and 1,10-phenanthroline inhibit their activity; III) they are secreted in a latent proform (zymogen) and hence require activation by limited proteolysis; IV) they are inhibited by specific tissue inhibitors or metalloproteinases (TIMPs). All of the enzymes have

61

now been cloned and show extensive sequence homology. They differ, however, in their substrate specificities. Interstitial collagenases cleave native fibrillar collagen at one specific locus per polypeptide chain. Gelatinases degrade denatured interstitial collagen and native basement membrane collagens, while stromelysins have a wide specificity against matrix components including gelatin, laminin, fibronectin, proteoglycan core proteins and elastin. In addition, the enzymes can be distinguished by the sizes of their latent and active forms (Table 1).

Emerging evidence that MMPs are involved in regulating VSMC proliferation and migration

Despite the possible importance of matrix remodelling for atherogenesis, the occurrence and function of MMPs in vascular cells has only recently been investigated. One report has described the production from calf aortic VSMC of interstitial collagenase (13). Endothelial cells are known to secrete gelatinases and TIMP (27, 28). Macrophages and some smooth muscle cells expressing stromelysin have been recently detected in human atherosclerotic plaques (11). While these observations provide some evidence for the secretion of MMPs by vascular cells, their role in VSMC proliferation and migration has not been tested directly.

To achieve this objective, we developed an experimental model to quantify the proliferation and migration of VSMC that are interacting with their native extracellular matrix. The model (29) consists of precise 1 mm square explants of rabbit aortic tunica media maintained individually in tissue culture. The rate of proliferation and outgrowth of cells is highly reproducible and all the outgrown cells are identified as VSMC by immunocytochemistry with anti-(smooth muscle actin). In the first part of our study (30), we took advantage of the availability of two structurally dissimilar, potent and specific synthetic MMP inhibitors (21). These agents incorporate a Zn^{2+} chelating group and a peptide analogue in their structures. They bind therefore to the active site of the MMPs and act as competitive inhibitors. Ro 31-4724 is a hydroxamic acid derivative which inhibits interstitial collagenase, gelatinase and stromelysin with

Table 1. The three classes of MMPs.

MMP	Mol. Weight in kDa		Substrates
	Latent	Active	
Interstitial Collagenases	55	45	Native fibrillar collagen. Gelatin (limited) Proteoglycan core protein (limited)
Type IV collagenases/gelatinases	72 95	66 88	Denatured interstitial collagens (gelatin) Elastin and fibronectin Basement membrane-type IV collagen
Stromelysins	57	48	Proteoglycan core protein. Elastin, fibronectin, and laminin. Gelatin (limited) (Activate procollagenases)

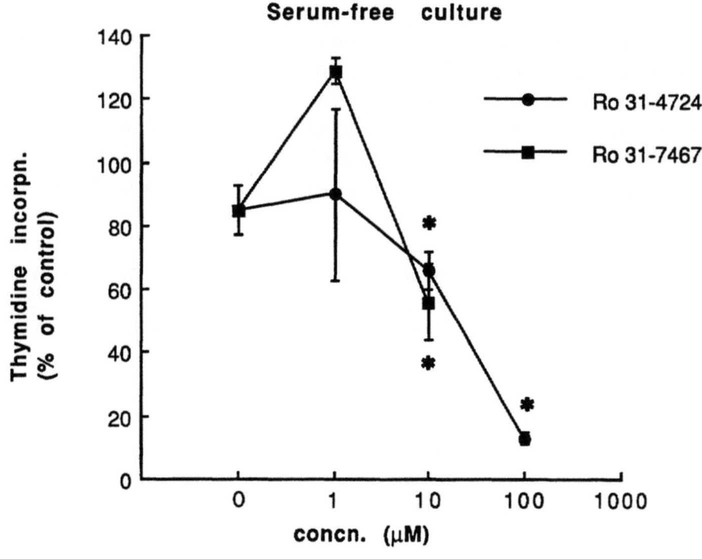

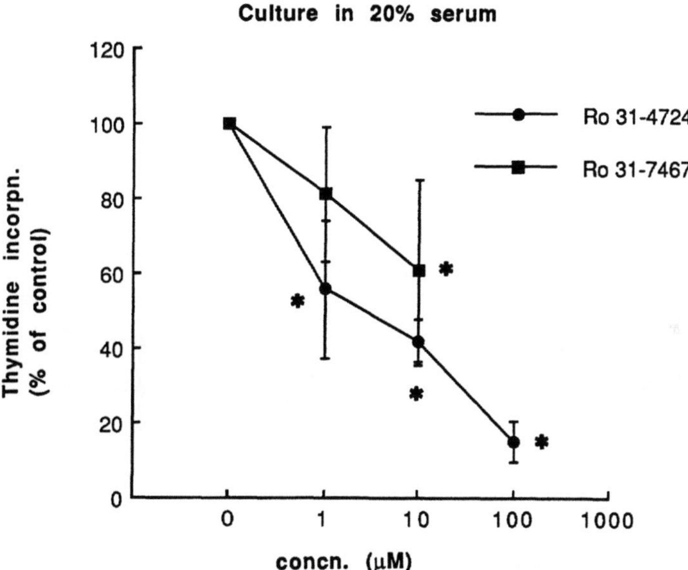

Fig. 3. Concentration-responses of proliferation to inhibitors. Aortic explants were cultured for 3 days in the absence (control) or presence of Ro 31-4724 (●) or Ro 31-7467 (■) at the final concentrations indicated in 0.5 % of DMSO. The incorporation of [^{3}H]thymidine (1 μCi/ml) over the last 18 h into trichloroacetic acid precipitates of eight extracts pooled to give three replicates for each condition was quantified and compared to that in control cultures.

a) The culture medium was DMEM containing 1 mg/ml of ovalbumin.

b) The culture medium was DMEM containing 20 % fetal bovine serum. Values are means ± SEM for three or four separate experiments.

* significantly (p < 0.05) less than 100 % (Data reproduced from reference [30] with permission.)

similar potency. Ro 31-7467 is a phosphinic acid analogue with similar potency against interstitial collagenase, but somewhat lower potency against gelatinase and stromelysin than Ro 31-4724 (21). Since their action involves Zn^{2+} binding, these agents do not inhibit the serine, cysteine or aspartate classes of proteinase.

Because serum is known to contain inhibitors of MMP, a series of experiments was first carried out in serum-free medium (Fig. 3a), taking advantage of the initial partial serum-independence of proliferation under these conditions. Ro 31-4724 or Ro 31–7467 inhibited proliferation in a concentration-dependent manner (Fig. 3a). Approximately 50 % inhibition occurred with 10 μM of Ro 31-4724 or Ro 31-7467. In a separate series of explants cultured in the presence of serum for 3 days, Ro 31-4724 or Ro 31-7467 also inhibited proliferation with a similar potency (Fig. 3b). Cytoxicity did not explain the inhibition of proliferation because neither MMP inhibitor reduced the concentration of ATP in explants (30). In a third series of experiments, the effects on proliferation of the two MMP inhibitors were measured in serum-containing medium at different time points. Both agents significantly inhibited [^{3}H]thymidine incorporation on day 3 and also at later time points (Fig. 4). By day 7, there was significant visible outgrowth of cells from the explants (Table 2). 100 μM of Ro 31-4724 significantly inhibited outgrowth, whereas 10 μM-Ro 31-7467 produced in inhibitory trend (Table 2).

In the second part of the study (30), we investigated whether VSMC secreted MMPs that were sensitive to the inhibitors. Conditioned media were obtained after

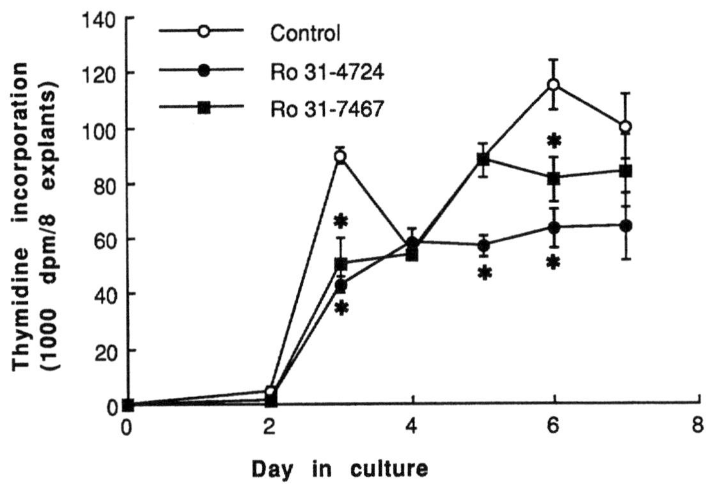

Fig. 4. Time-course of the effect of MMP inhibitors on proliferation. Aortic explants were incubated for the times shown with 0.5 % of DMSO (vehicle, O) or 0.5 % of DMSO and 100 μM-Ro 31-4724 (●) or 10 μM-Ro 31-7467 (■); incorporation of [^{3}H]thymidine (1 μCi/ml) was measured over the last 18 h. Values are means ± SEM (when larger than the symbol) from three separate experiments, each performed in triplicate.

* p < 0.05 vs. 0.5 % of DMSO on the same day (Data reproduced from reference [30] with permission.)

Table 2. Outgrowth of rabbit aortic smooth muscle cells

Condition	Visually observed coverage on day 7		
	none	0–5 %	5–50 %
Vehicle control	3 ± 1	6 ± 5	15 ± 3
Ro 31-4724 (100 μM)	12 ± 2*	8 ± 3	4 ± 2*
Ro 31-7467 (10 μM)	8 ± 2	6 ± 1	10 ± 3

24 wells were scored in each of three different preparations. The data show the number of wells with the degrees of outgrowth specified.
* $p < 0.05$ vs control (Student's *t*-test)

3 days in serum-free culture from the same explants used for the growth studies. When these were subjected to zymography in gels impregnated with gelatin, a 95 kDa progelatinase and a 72 kDa progelatinases together with its 68 kDa active form were detected (Fig. 5). The enzymes were definitively identified (30) using specific antisera to the purified pig 95 kDa and human 72 kDa gelatinases. Extracts of the explants themselves showed a similar, although much weaker pattern (30), indicating that the enzymes were mainly secreted. In support of this, similar zymograms were

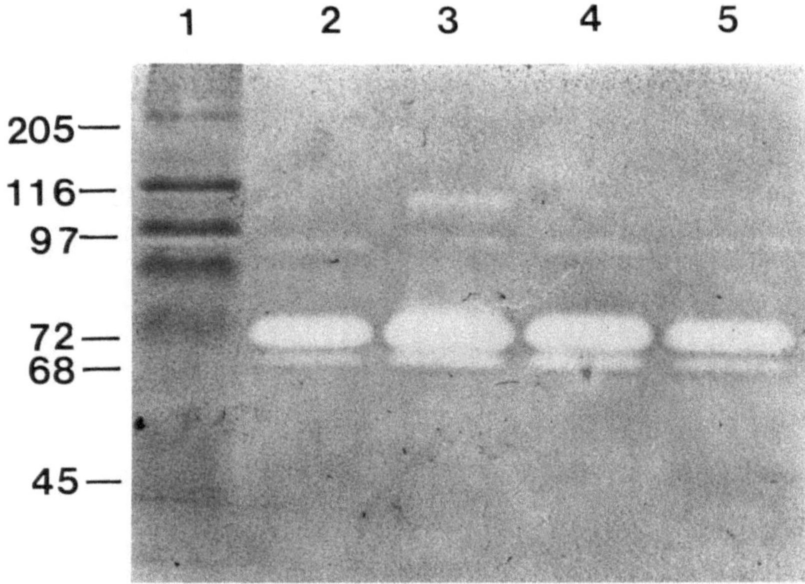

Fig. 5. MMP activity in serum-free culture. The medium from the rabbit aortic explants described in Fig. 3a were concentrated 10-fold and subjected to SDS-gel electrophoresis into gels impregnated with gelatin. After removing SDS, the gels were incubated overnight at 37 °C and then stained with Coomassie blue. The presence of enzymes was detected as zones of lysis of gelatin. Samples comprised: Lane 1) molecular weight markers; 3) to 6) conditioned media of explants from different aortas.

65

obtained from media removed repeatedly at 2-day intervals from the same culture (30). The gelatinase activities detected by zymography were inhibited in a concentration-dependent manner by Ro 31-4724 (Fig. 6a) or Ro 31-7467 (Fig. 6b). The activities were almost abolished by 10 μM of Ro 31-4724 or 50 μM of Ro 31-7467. Stromelysin was not detected in the conditioned medium by zymography or Western blotting nor was interstitial collagenase by functional assay or Western blotting (30).

Implications

These data directly implicate MMPs in the proliferation and migration of VSMC. MMP inhibitors were most effective in the first 3 days of culture, which implies that MMPs were involved in the remodelling of the native extracellular matrix within the explants. MMPs also appeared to be involved in permitting migration of cells from the explants at later times. Whether the later effects on proliferation represent carry-over from early inhibition or whether the inhibitors acted also on outgrown cells that were elaborating their own extracellular matrix, we do not yet know.

Detection of gelatinase enzymes but not interstitial collagenase or stromelysins implies that inhibition of gelatinase was the basis for the effects of drugs. However, because the inhibitors have a broad specificity for MMPs, we cannot rule out the involvement of the other enzymes. The structural dissimilarity of the compounds makes it unlikely that a mechanism of action other than inhibition of MMPs accounts for their observed effect on proliferation. In further support, the concentration responses for inhibition of gelatinase activity (Fig. 6) and proliferation (Fig. 3) were similar. The most direct interpretation of our results is therefore that gelatinase activity was required (probably to degrade the pericellular basement membrane) in order for cells to proliferate and migrate from the explants. The biochemical basis for this requirement remains to be elucidated. As discussed above, possibilities include the removal of a physical constraint to migration, relief of direct inhibitory interactions with basement membrane components or the release of latent growth factors.

The role for MMPs implied here in the context of atherosclerosis is similar to that established on the basis of more exhaustive studies in other cellular pathologies, including glomerularsclerosis, rheumatoid arthritis and tumour metastasis (5, 14, 20, 34). Moreover, it provides a possible explanation for the strong association between aortic atherosclerosis and aneurysm formation (30), long recognised as a disease of excessive matrix degradation. It invites a further speculation that inappropriate activation of MMPs may be a common molecular mechanisms underlying plaque rupture and aneurysm formation. Although we have emphasised the role of MMPs derived from VSMC, in the intact vessel, other cellular sources may contribute. As detailed above, fibroblasts and endothelial cells may also secrete latent forms of MMPs as well as TIMP (27, 28). Recent in situ hybridization studies have suggested that macrophages in atherosclerotic plaques are also a potential source of stromelysin (11).

The ability to maintain low rates of VSMC proliferation in normal vessels while sustaining turnover of all matrix components implies that MMP activity is tightly regulated and prompts a search for the mechanisms involved. Despite the paucity of information directly relating to the blood vessel wall, there is considerable relevant information from other cell types. Three levels of control are currently recognised (5, 20, 34), namely regulation of expression of the MMP genes, proteolytic activation of the latent proenzymes and regulation of the expression of genes for specific TIMPs.

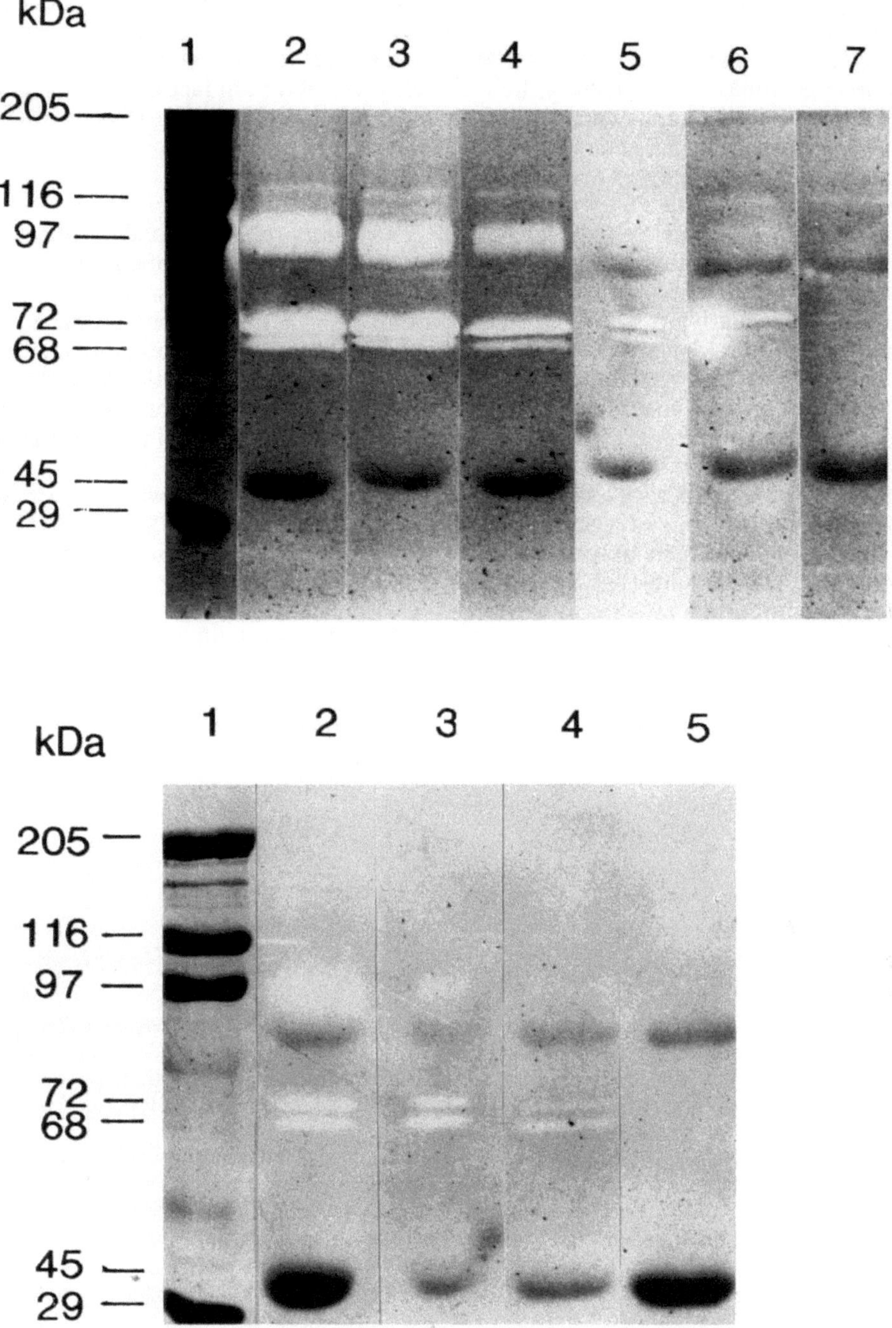

Fig. 6. Effect of MMP inhibitors on gelatinase activity. Conditioned media were subjected to zymography in buffer containing 0.5% DMSO (vehicle) and the concentrations of MMP inhibitors as follows: Panel a) 0, 0.1, 0.5, 1, 10 and 100 μM-Ro 31-4724; Panel b) 0, 1, 10 and 50 μM-Ro 31-7467. (Data reproduced from reference [30] with permission.)

The synthesis and secretion of the 95 kDa progelatinase is induced by cytokines (e.g. IL-1ß and TNFα), polypeptide growth factors (e.g. PDGF, bFGF) and glucocorticoids, as well as phorbol esters (32). Indeed, the promotor region of the gene for the 95 kDa gelatinase has a binding site for nuclear transcription factors (the so-called TPA responsive element or activator protein-1 binding site) (12). The structure of the promotor region of the gene for the 72 kDa gelatinase suggests it is under different control and may be largely constitutive, as suggested by our results.

Control of MMP activity at the level of conversion of the proenzymes to their active forms is best understood for procollagenase but probably applies to the other MMPs. Limited proteolysis by trypsin-like proteases, cathepsins B and L, PMN elastase and plasmin could represent physiologically relevant pathways. A role for plasmin is particularly attractive because plasminogen activators are produced by endothelial cells, leucocytes and fibroblasts (32). Moreover, Clowes et al. have recently shown that VSMC themselves express uPA during proliferation and tPA during migration (4). Thus the conditions appear right in the microenvironment surrounding the proliferating VSMC for activation of latent progelatinase, consistent with our observation of the active 68 kDa gelatinase in Fig. 5 and 6.

Conversely, inhibitory control of MMPs occurs by interaction with at least two isoforms of TIMP, one of which has been shown to be secreted by VSMC (13). However, few studies in any tissue have investigated the balance of expression between MMPs and TIMPs, although they appear not to be induced in parallel (17). This is clearly an important area for future research.

In considering possible future developments in therapy, inhibitors of MMPs may provide an exciting new approach. If MMPs are involved in plaque rupture, such inhibitors might have the advantage of stabilising existing lesions as well as reducing progression. Inhibitors of MMPs are currently in clinical trial for metastasis of breast tumours, so they may soon be available for evaluation in models of human vascular disease.

References

1. Burke JM, Ross R (1979) Synthesis of connective tissue macromolecules by smooth muscle. Int Rev Connect Tissue Res 8: 119–157
2. Campbell JH, Black MJ, Campbell GR (1989) Replication of smooth muscle cells in atherosclerosis and hypertension. In: Meyer P, Marche P ed. Hypertension and Atherosclerosis. New York: Raven Press, 15–33
3. Clowes AW, Reidy MA, Clowes MM (1983) Kinetics of cellular proliferation after arterial injury. Smooth muscle growth in the absence of endothelium. Lab Invest 49: 327–333
4. Clowes AW, Clowes MM, Reidy MA. Belin D (1990) Smooth muscle cells express urokinase during mitogenesis and tissue-type plasminogen activator during migration in injured rat carotid artery. Circ Res 67: 61–67
5. Davies M, Martin J, Thomas GT, Lovett D (1992) Proteinases and glomerular matrix turnover. Kidney Int 41: 671–678
6. Faggiotto A, Ross R, Harker L (1984) Studies of hypercholesterolaemia in the non-human primate. 1. Changes that lead to fatty streak formation. Arteriosclerosis 4: 323–340
7. Ferns GAA, Raines EW, Sprugel KH, Motani AS, Reidy MA, Ross R (1991) Inhibition of neointimal smooth muscle accumulation after angioplasty by an antibody to PDGF. Science 253: 1129–1132

8. Groves PH, Lewis MJ, Newby AC, Cheadle HA, Penny WJ (1992) Progressive intimal thickening after balloon angioplasty is associated with rupture of the internal elasic lamina. Br H Journal 68: 86

9. Hangartner JRW, Charleston AJ, Davies MJ, Thomas AC (1986) Morphological characteristics of clinically significant coronary artery stenosis in stable angina. Br Heart J 56: 501–508

10. Hedin U, Bottger BA, Forsberg E, Johansson S, Thyberg J (1988) Diverse effects of fibronectin and laminin on phenotypic properties of cultured smooth muscle cells. J Cell Biol 107: 307–319

11. Henney AM, Wakely PR, Davies MJ et al. (1991) Localization of stromelysin gene expression in atherosclerotic plaques by in situ hybridization. Proc Natl Acad Sci USA 88 8154–8158

12. Huhtala P, Tuuttila A, Chow LT, Lohi J, Keski-Oja J, Tryggvason K (1991) Complete structure of human gene for 92-kDa type IV collagenase: Divergent regulation of expression for the 92- and 72-kilodalton enzyme genes in HT-1080 cells. J Biol Chem 266: 16485–16490

13. Kishi J, Hayakawa T (1989) Synthesis of latent collagenase and collagenase inhibitor by bovine aortic medial explants and cultured medial smooth muscle cells. Connective Tissue Res 19: 63–76

14. Liotta LA, Steeg PS, Stetler-Stevenson WG (1991) Cancer metastasis and angiogenesis: an imbalance of positive and negative regulation. Cell 64: 327–336

15. Majack RA, Cook SC, Bornstein P (1986) Control of smooth muscle cell growth by components of the extracellular matrix: autocrine role for thrombospondin. Proc Natl Acad Sci USA 83: 9050–9054

16. Masuda J, Ross R (1990) Atherogenesis During Low Level Hypercholesterolemia in the Nonhuman Primate. 1. Fatty Streak Formation. Arteriosclerosis 10: 164–177

17. Matrisian LM (1990) Metalloproteinases and their inhibitors in matrix remodelling. Trends Gerontol 6: 121–125

18. McCaffrey TA, Falcone DJ, Brayton CF, Agarwal LA, Welt FGP, Weksler BB (1989) Transforming growth factor-β activity is potentiated by heparin via dissociation of the transforming growth factor-β/alpha2-macroglobulin complex. J Cell Biol 109: 441–448

19. Moczar M, Lafuma C (1986) Structural glycoproteins from aorta and lung. Frontiers Matrix Biol 11: 42–57

20. Murphy G, Docherty AJP, Hembry RM, Reynolds JJ (1991) Metalloproteinases and tissue damage. Br J Rheumatol 30 (Suppl 1): 25–31

21. Nixon JS, Bottomley KMK, Broadhurst MJ et al. (1991) Potent collagenase inhibitors prevent IL-1 induced cartilage degradation in vitro. Int J Tissue React 13: 237–243

22. Orekhov AN, Karpova II, Tertov VV et al. (1984) Cellular composition of atherosclerotic and uninvolved human aortic subendothelial intima: Light microscope study of dissociated aortic cells. Am J Pathol 115: 17–24

23. Reed D, Reed C, Stemmermann G, Hayashi T (1992) Are aortic aneurysms caused by atherosclerosis? Circulation 85: 205–211

24. Reidy MA, Silver M (1985) Endothelial regeneration – Lack of intimal proliferation after defined injury to rat aortas. Am J Pathol 118: 173–177

25. Ross R (1986) The pathogenesis of atherosclerosis. An update. N Engl J Med 314: 488–500

26. Saksela O, Rifkin DB (1990) Release of basic fibroblast growth factor-heparan sulfate complexes from endothelial cells by plasminogen activator-mediated proteolytic activity. J Cell Biol 110: 767–775

27. Scott-Herron G, Werb Z, Dwyer K, Banda MJ (1986) Secretion of Metalloproteinases by Stimulated Capillary Endothelial Cells. 1 Production of Procollagenase and Prostomelysin Exceeds Expression of Proteolytic Activity. J Biol Chem 261: 2810–2813

28. Scott-Herron G, Banda J, Clark J, Gavrilovic J, Werb Z (1986) Secretion of Metalloproteinases by Stimulated Capillary Endothelial Cells. 2. Expression of Collagenase and Stromelysin Activities in Regulated by Endogenous Inhibitors. J Biol Chem 261: 2814–2818

29. Southgate KM, Newby AC (1990) Serum-induced proliferation of rabbit aortic smooth muscle cells from the contractile state is inhibited by 8-Br-cAMP but not 8-Br-cGMP. Atherosclerosis 82: 113–123

30. Southgate KM, Davies M, Booth RFG, Newby AC (1992) Involvement of extracellular matrix degrading metalloproteinases in rabbit aortic smooth muscle cell proliferation. Biochem J 288: 93–99
31. Thyberg J, Hedin U, Sjolund M, Palmberg L, Bottger BA (1990) Regulation of differentiated properties and proliferation of arterial smooth muscle cells. Arteriosclerosis 10: 966–990
32. Werb Z (1989) Proteinases and matrix degradation. In: Kelley WN Jnr HED, Ruddy S, Sledge CB ed. Textbook of Rheumatology. Philadelphia: W B Saunders and Co, 300–321
33. Wight TN (1990) The cell biology of arterial proteoglycans. Arteriosclerosis 9: 1–20
34. Woessner JFJ (1991) Matrix metalloproteinases and their inhibitors in connective tissue remodeling. FASEB J 5: 2145–2154

Author's address:
Andrew C. Newby, MD
Department of Cardiology
University of Wales College of Medicine
Heath Park
Cardiff, CF4 4XN
UK

Vascular renin-angiotensin-system, endothelial function and atherosclerosis?

J. Holtz, R. M. Goetz

Institut für Pathophysiologie Martin-Luther-Universität Halle-Wittenberg

Summary: Clinical observations demonstrate an enhanced risk for myocardial infarction in patients with substained activation of the local and/or systemic renin-angiotensin system, such as a high renin-sodium profile or a heritably enhanced expression of angiotensin converting enzyme. Chronic renin-angiotensin system blockade by angiotensin converting enzyme inhibition in patients with moderate heart failure reduces the rate of myocardial infarction and reinfarction. Preliminary experimental evidence suggests that these clinical observations may be partially explained by a proatherogenic effect of an activated renin-angiotensin system, which can downregulate the endothelial releasability of nitric oxide. Nitric oxide exerts many potentially antiatherogenic effects on endothelium, platelets and low density lipoproteins and indirectly on monocytes and leukocytes. Hypertension-induced chronic distension of elastic arteries upregulates the local renin-angiotensin system in these arteries and thereby downregulates nitric oxide releasability. Enhanced local synthesis of the trophic factor angiotensin-II and reduced releasability of the antitrophic factor nitric oxide appear to cooperate in the trophic adaptation of the distended vessel wall to the enhanced load, but with the disadvantage of enhanced susceptibility for atheroma development due to reduced releasability of nitric oxide. Chronic blockade of the renin angiotensin system by angiotensin converting enzyme inhibitors or by angiotensin receptor type-1 antagonists normalizes a reduced endothelial releasability of nitric oxide in several models, partially by a bradykinin-dependent mechanism. This endothelial protection proved to attenuate the progression of atheroslerosis in experimental models. The antiatherogenic potential of renin angiotensin system blockade in humans is presently under study.

Key words: Nitric oxide – monocyte adhesion – platelet activation – ACE inhibition – angiotensin receptor antagonist – endothelial protection

Introduction

The association in the title of this text cannot yet be regarded as an established fact, the question mark is an essential part of the title and should not be overlooked! Several experimental data in support of this putative association are now available and will be discussed here briefly. Altogether, these experimental data do not yet yield definitive proof for the association of the title. However, they do nourish the expectation that chronic pharmacological interference with the local and/or systemic renin-angiotensin system might contain an antiatherosclerotic potential. Such a potential could be of paramount importance for prevention of cardiovascular morbidity and mortality. Therefore, it will require intensive clinical research and critical testing patients in order to delineate the practical relevance of such a hitherto speculative hypothesis. Interestingly, however, several recent clinical observations, partially unexpected and incompletely understood, pushed this speculative association between the renin-angiotensin system and atherosclerosis to the awareness of the medical community (Tab. 1).

More than a decade ago, the group of John Laragh at Cornell University used the "renin-sodium profile" in hypertensives (i.e.: the relation of 24-h sodium excretion to plasma renin activity) as an approach to understand pathophysiologic aspects of the altered pressure and volume homeostasis in these patients (95). Later, a long-term prospective study demonstrated that an elevated renin-sodium profile is an independent risk factor for myocardial infarction even after adjustment for other risk modulators including blood pressure, history of cardiovascular treatment or use of β-blocker (2). Last year, Soubrier and colleagues (19) published their spectacular observation on an association between myocardial infarction and the deletion polymorphism in the noncoding region of the gene for angiotensin converting enzyme (ACE). Finally, two prospective, randomized long-term studies with ACE inhibitor treatment in patients with moderate heart failure demonstrated a reduction in the rate of myocardial reinfarctions or infarctions under chronic ACE inhibition (84, 100). This reduction is not easily explained by vasodilation and attenuation of neuroendocrine activity alone and may indicate a new, hitherto unidentified preventive potential of these drugs (27).

The mechanisms involved in these observations have not yet been identified, and it has not been demonstrated for any of these observations that the alterations in the rate of myocardial infarctions were due to alterations in the development or progression of atherosclerosis. Nevertheless, together with the recent demonstration that circulating renin is taken up into vascular tissue and does contribute to local vascular angiotensin production in humans (102), these observations are consistent with speculative associations between the local activity of the renin-angiotensin system and the susceptibility for atherogenesis (Table 1). Experimental data, to be discussed in the next chapters of this text, indicate that the local availability of endothelium-derived relaxing factor (EDRF) might be a critical link in this speculative association.

Table 1. Associations of vascular renin-angiotensin-system and myocardial infarction in humans

Observation	Hypothetic mechanisms
High renin-sodium profile in hypertensives is independent risk factor for myocardial infarction (2)	Circulating renin contributing to local vascular angiotensin production after uptake into vessels; enhanced local vascular renin-angiotensin formation atherogenic?
A deletion polymorphism in the non-coding region of the ACE gene is a potent risk factor for myocardial infarction and associated with elevated plasma ACE (3)	Deletion polymorphism associated with enhanced vascular ACE expression and activity? Enhanced expression of elements of the vascular renin-angiotensin system atherogenic?
In patients with moderate heart failure, prolonged treatment with ACE inhibitors reduced the rate of myocardial reinfarctions or first infarctions (4, 5)	Preventive effect of ACE-inhibitors due to reduction in development/progression of atherosclerotic lesions by reducing local vascular ACE activity?

72

Whether this is true for humans and whether this is the explanation for the clinical observations summarized in Table 1, is entirely unproven and remains to be demonstrated.

Endothelial nitric oxide deficiency and atherogenic mechanisms

Present concepts on the genesis of atherosclerosis assume functional alterations of morphologically intact endothelial cells as early, initiating events, while classical concepts postulated endothelial denudation as the primary step (89). An important role in the development of atherosclerotic lesions is ascribed to the invasion of blood-derived monocytes to vessels with morphologically intact endothelium (42, 43, 101). The adhesion of circulating monocytes to the endothelium and their diapedesis between endothelial cells into the subendothelial space is the precondition for their subsequent conversion into macrophages, which begin to accumulate lipids and take on the characteristics of foam cells, the major constituent of the fatty streaks (42, 43, 101). This adhesion of monocytes is facilitated by the endothelial synthesis of monocyte chemotactic protein 1 (MCP-1), an important promoter of monocyte accumulation, and the endothelial synthesis and expression of this chemoattractant factor is stimulated by exposure of endothelial cells to minimally oxidatively modified low-density lipoproteins (m-oxLDL, 10, 29, 90).

The enhanced synthesis of endothelial MCP-1 by m-oxLDL is probably mediated by suppression or inactivation of EDRF with its active principle nitric oxide (90). Nitric oxide is not only a paracrine dilator of vascular smooth muscle and inhibitor of platelet activation (88), but also an autocrine modulator of endothelial synthesis of signal molecules: suppression of nitric oxide synthesis by inhibitory derivatives of L-arginine enhances MCP-1 expression, while addition of the nictric oxide precursor L-arginine reduces the MCP-1 expression (90). Endothelium-dependent, nitric oxide-mediated vasodilation is attenuated by m-oxLDL (40, 104, 107), this attenuation is mediated partially by the inactivation of EDRF (23) and can be reversed in vitro by high density lipoproteins (73). The factors stimulating the expression and synthesis of endothelial MCP-1 are not yet completely identified, but the endothelium-derived nitric oxide obviously exerts an autocrine inhibitory influence on MCP-1 expression (90).

Probably, the enhancement of vascular monocyte accumulation by hypercholesterinemia and m-oxLDL is not only brought about by the EDRF-modulated endothelial synthesis of the chemoattractant MCP-1. Adhesion molecules mediate many kinds of cell-to-cell-interactions, including adherence of leucocytes and their infiltration during inflammation as well as adherence and aggregation of platelets during hemostasis and thrombosis. A specific endothelial cell adhesion molecule involved in adhesion of monocytes to endothelial cells has been identified recently (30). The expression of this endothelial adhesion protein can be induced by endotoxin. It contributes to monocyte recruitment in hypercholesteremia, since its endothelial expression is enhanced in hypercholesteremic rabbits, specifically in aortic segments with developing intimal lesions (30). The mechanisms causing the upregulation of this monocyte binding endothelial cell adhesion molecule in hypercholesterinemia have not yet been identified. However, it is tempting to speculate that the suppression of endothelial availability of nitric oxide is involved in this upregulation: the endothelial expression of fibronectin (another adhesion molecule involved in the adhesion of collagen and fibrin to cell membranes) is enhanced by suppression of endothelial synthesis of nitric oxide and is inhibited by activation of nitric oxide release (90). The autocrine modulation of endothelial expression of MCP-1 and of the monocyte cell adhesion molecule may explain the previous observation that nitric oxide is an inhibitor of monocyte adherence to activated endothelium in vitro and of monocyte migration induced by the chemotactic formyl-methionyl-leucyl-phenyl-alanine (5). Any attenuation of EDRF-availability, induced by m-ox-LDL or other impairments of endothelial nitric oxide release, should enhance the synthesis of MCP-1 and of monocyte binding adhesion molecules, thereby augmenting monocyte accumulation in the vascular wall. On the other hand, endothelium-derived nitric oxide has been shown to partially protect LDL-particles against oxidative modification by macrophages in vitro (71). The relevance of this observation under in vivo conditions is difficult to assess, but any local reduction in endothelial availability might facilitate monocyte adhesion and their conversion to macrophages, which in turn may allow further oxidative modification of LDL and further suppression of local EDRF availability.

Endothelium-derived nitric oxide modulates also another interaction of circulation blood cells with damaged or altered tissue via leukocyte adhesion molecules of the integrin class. Integrins are a class of heterodimeric membrane glycoproteins, consisting of noncovalently associated α- and β-subunits, and are involved in adhesion of cells to molecules of the extracellular matrix and to other cells (55). Integrins of the CD11/CD18 class mediate binding of fibrinogen, coagulation factors, endotoxin and complement components to activated neutrophils and macrophages. The size of experimental myocardial infarctions after ischemia and reperfusion can be reduced by antibodies against CD11/CD18 via attenuating neutrophil adhesive interactions (98, 99). The application of nitric oxide synthase inhibitors is associated with a substantially enhanced leukocyte adherence and infiltration into postischemic rat mesenteric vessels, and this enhancement is due to upregulation of the adhesive integrin CD11/CD18 and can be attenuated by the administration of the nitric oxide precursor L-arginine (61). Similarly, exogenous nitric oxide (applied as infusion of sodium nitrite) can reduce neutrophil adherence and accumulation and attenuate myocardial and endothelial damage in postischemic, reperfused myocardium, and this protection is probably not only due to the dilatory action of the nitrite (59). The modulation of leukocyte adherence by endogenous or exogenous nitric oxide has been analyzed in models of postischemic reperfusion, but neutrophil adherence to functionally altered endothelium may also contribute to atherogenic processes in hypercholesterinemia (63, 64). Therefore, a reduction in endothelial releasability of nitric oxide might also enhance the susceptibility of atherosclerotic lesion development by enhancing leukocyte adherence and activation.

The adhesion and aggregation of platelets to the altered luminal surface of fatty streaks and ulcerating lesions is an important mechanism in the progression of atherosclerosis and contributes to the most dangerous arteriosclerosis-associated complication, the intravasal thrombosis (89). Platelets have a high content of soluble guanylate cyclase, the physiological target of endothelium-derived nitric oxide (12), and endothelium-derived relaxant factor elevates the cGMP content of platelets and thereby inhibits the activation of platelets during their contact to the luminal surface of functionally intact endothelial cells in vitro and in vivo (4, 13, 39, 50, 54, 88). In this inhibition of platelet activation, endothelium-derived relaxing factor acts synergistically to prostacyclin (88), which elevates cAMP content of platelets. Bioassay experiments in atherosclerotic arteries demonstrate attenuated release of endothelium-derived relaxing factor (48), and this is due to inactivation of nitric oxide by enhanced production of superoxide anion (47, 51, 52, 72, 103). The overall production of nitric oxide appears to be augmented in atheroslerotic vessels (75), probably not due to physiological endothelial nitric oxide formation by the constitutive nitric oxide synthase of endothelial cells, but rather due to the inflammation-related nitric oxide formation by the cytokine inducible nitric oxide synthase, which can be expressed in many cell types during inflammation. The depressed endothelial nitric oxide formation in early stages of atheroma development as well as the enhanced inactivation of nitric oxide in vessels with advanced arteriosclerosis may facilitate platelet activation and augment the contribution of platelet activation to the atherosclerotic process.

Consistent with these speculative considerations is the recent documentation that dietary supplementation of the nitric oxide precursor L-arginine in hypercholesteremic rabbits has an anti-atherogenic effect (28). The affinity of the various nitric oxide synthases to L-arginine is different and it is not quite clear how this dietary supplementation of L-arginine really enhances the local nitric oxide availability in atherosclerosis-prone arteries. However, the observation is an important argument in favor of the hypothesis that reduced local endothelial availablity can contribute to atheroma development and progression. Table 2 summarizes the experimental findings supporting this hypothesis.

Hypertension-induced attenuation of endothelial nitric oxide releasability

An attenuation of endothelium-mediated vasodilation has been observed in large conductance vessels of animals with experimental or genetic hypertension (67, 68,

Table 2. Antiatherogenic efficacy of endothelial nitric oxide?

– Nitric oxide inhibits expression of monocyte chemoattractant protein (MCP-1) in endothelial cells in vitro (14)

– Nitric oxide inhibits monocyte adhesion to activated endothelium in vitro and FMLP-induced monocyte migration in vitro (22)

– Nitric oxide inhibits platelet activation and adhesion to endothelium in vitro and in vivo (15, 32–36)

– Nitrovasodilators reduce myocardial leukocyte adhesion and accumulation in postischemic reperfusion by a mechanism not related to the vasodilation (28)

– Leukocyte adhesion in postischemic mesenteric vessels is potentiated by NO-synthase antagonists (via upregulation of CD_{11}/CD_{18}) and attenuated by the nitric oxide precursor L-arginine (27)

– Nitric oxide protects LDL against oxidative modification by macrophages in vitro (23)

– Dietary supplementation of nitric oxide precursor L-arginine attenuates progression of atherosclerosis in hypercholesteremic rabbits (44)

113), in resistance vessels of hypertensive animals (33, 34, 105) and in the vasculature of humans with essential hypertension (18, 65, 82, 83). Such alterations in endothelial control of vascular tone could result mainly from dysfunctions of the dilatory endothelial L-arginine/nitric oxide pathway and/or from enhanced formation of endothelium-derived contracting factors such as prostaglandin H_2 or endothelin (69). The contribution of these alterations appears to vary, depending on the type of hypertension and the vessel under study, but an attenuated function of the endothelial nitric oxide pathway always contributes to some degree to the endothelial dysfunction in hypertension. This attenuation of nitric oxide mediated dilation in hypertension is probably not due to a reduced responsiveness of vascular smooth muscle to nitric oxide or due to the increased formation of the endogenous inhibitor of nitric oxide synthase (70, 108). However, altered expression of endothelial receptors or their signal transduction elements, decreased endothelial nitric oxide synthase or increased breakdown of the formed nitric oxide all could be involved to a variable degree (69).

In rats with hypertension induced by aortic coarctation, this depression of endothelium-dependent, nitric oxide-mediated dilation of large distensible arteries is strictly localized, restricted only to the vessels exposed to the enhanced blood pressure (44, 45). This depression is not explained by altered responsiveness of the vasculature to nitric oxide, is resistant to cyclooxygenase inhibition, and can be best explained by a reduced endothelial releasability of nitric oxide (44, 45). In such a model of reactive hypertension, a rather sudden distension of the prestenotic arteries takes place and should trigger trophic responses of the distended arterial wall as adaptation to the altered load. It is tempting to speculate that the blunted releasability of nitric oxide from the endothelium of those distended vessels facilitates such a trophic adaptive response, since nitric oxide can act as inhibitor of smooth muscle proliferation (3, 78, 93). This inhibition is, at least partially, mediated by the nitric oxide-induced formation of intracellular cGMP (41, 56, 57, 58).

On the other hand, trophic responses of the vascular wall in response to altered hemodynamic load have been associated with activation of the local renin-angiotensin system, since angiotensin can be regarded as a paracrine and autocrine trophic factor or cofactor in the vasculature (92). The fascinating question is whether the suppression of the antitrophic signal nitric oxide and the activation of the trophic factor angiotensin are coordinated reactions of arteries which regulate trophic adaptations in response to enhanced chronic distension.

**Vascular renin-angiotensin system:
modulator of endothelial releasability of nitric oxide?**

This question came up first from observations on endthelium-mediated dilations of aortic rings in vitro from hypertensive rats treated by different antihypertensive drugs (26, 74). Spontaneously hypertensive rats treated by ACE inhibitors had a better endothelium-mediated dilatory responsiveness than rats treated by hydralazine (26). Reversal of experimental hypertension improves endothelium-dependent aortic relaxation in high renin and low renin models of hypertension (67). However, the difference in endothelial function after treatment by hydralazin or by ACE inhibition (26) could not be explained by this hypotension-induced normalization, since the lowering of blood pressure was similar with both treatments (26). Furthermore, chronic ACE inhibition in normotensive rats also augmented endothelium-mediated dilation without measurably lowering blood pressure (97), and acute ACE inhibition in patients with essential hypertension improved endothelium-mediated forearm vasodilation to acetylcholine by a mechanism which was not due to enhanced responsiveness of vascular smooth muscles to nitrovasodilators, and which could not be explained by the acute hypotension (53).

An inverse relation of local vascular ACE activity and endothelial releasability of nitric oxide in distensible arteries was demonstrated recently in rats with hypertension due to suprarenal coarctation of the abdominal aorta (44, 45). In this model, local ACE activity and ACE mRNA expression is elevated in elastic arteries (44, 45), as has been demonstrated by others in several models of renal or genetic hypertension (76, 77, 79, 80, 96). However, by placing the coarctation at various sites in the coarctation model, it was shown that this upregulation is restricted to the distensible arteries exposed to the hypertension (i.e.: proximal to the coarctation), and is not observed in elastic arteries distal to the coarctation or in the microvasculature of organs supplied by arteries originating proximal or distal from the coarctation [(44, 45), and unpublished observations from our laboratory]. A depression of endothelium-dependent, nitric oxide-mediated aortic dilatation to acetylcholine or histamine in vitro was only observed in the arteries with elevated ACE activity, which were exposed to the chronic distension (44, 45), but not in the arteries distal to the coarctation, in which ACE was normal. We treated hypertensive rats with coarctation hypertension by ACE inhibition for 2 weeks with a low dose, which did not modify the hypertension. This treatment lowered local ACE activity in the hypertension-exposed aortic segments below control levels and partially normalized the nitric oxide-mediated dilations without affecting endothelium-independent dilations to sodium-nitroprussid (44, 45). These observations indicated that the distension-induced upregulation of local ACE activity in elastic arteries is causal in downregulating local nitric oxide releasability in response to stimuli, which are not degradated by ACE activity. Interestingly, the restoration of endothelial nitric oxide releasability by chronic, nonhypotensive ACE inhibition was prevented by chronic coadministration of the bradykinin-2 receptor antagonist HOE 140 (44, 45). From these observations we propose a hypothesis on the interaction of local ACE and endothelial function during the adaptive phase in chronically distended elastic arteries, as outlined in Fig. 1.

Antiatherosclerotic effects by therapeutic suppression of the renin-angiotensin system via endothelial protection?

The central element in our hypothesis is the assumption of a distension-induced upregulation of ACE-inhibition in elastic arteries (Fig. 1). We have not yet quantified the degree of aortic distension in the coarctation model, and a cellular model for distension-induced ACE expression in endothelial cells or in vascular smooth muscle is not available. Furthermore, it is conceivable that the upregulation of ACE expression is induced by another unidentified trophic signal originating in the distended elastic arteries. Thus, this part of our hypothesis can be regarded as a plausible speculation explaining our observations. An upregulated ACE increases the capacity for

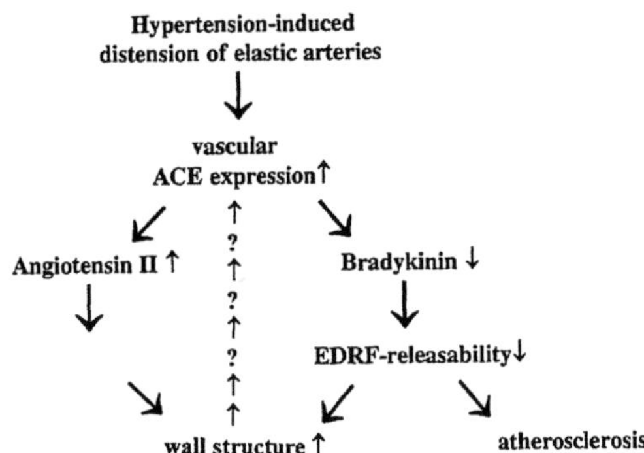

Fig. 1. Hypothesis connecting distension-induced upregulation of ACE expression in elastic arteries, the trophic adaptation of the vascular wall to the chronic distension and the susceptibility for atherosclerosis due to reduced releasability of EDRF/nitric oxide. The enhanced formation of the trophic factor angiotensin and the reduced releasability of the antitrophic signal nitric oxide cooperate in mediating the trophic adaptation of the vessel wall structure. Since chronic, nonhypotensive AT_1-blockade normalizes the elevated vascular ACE activity, a hypothetical positive feedback from the vascular wall under trophic stimulation to vascular/endothelial ACE expression is proposed.

local production of angiotensin-II, which is further enhanced by augmented expression of angiotensinogen in the distended arteries (96). The augmented local production of angiotensin-II could contribute to the trophic adaptation of the vessel wall to chronic mechanical distension. The enhanced local angiotensin-II production has not been assessed directly in the model, nor is the putative trophic role of locally formed angiotensin quantified.

The other consequence of the upregulated ACE in our hypothetic scheme deals with local bradykinin degradation (Fig. 1). Provided that local formation of bradykinin is not altered, any change in the local ACE activity should induce inverse changes in local bradykinin availability (14, 112), since it is degraded in part by this enzyme (36). In fact, many effects of ACE are regarded as bradykinin-mediated via the release of prostacyclin and nitric oxide, since they are attenuated by kinin antibodies, bradykinin antagonists or bradykinin deficiency (8, 9, 16, 17, 21, 22, 31, 49, 66, 87). In our hypothetic scheme, we propose that the chronically elevated ACE reduces bradykinin availability and thereby reduces the releasability of endothelium-derived nitric oxide (Fig. 1), since the improvement of nitric oxide releasability by chronic ACE inhibition was abolished by the bradykinin antagonist HOE 140 (44, 45).

This proposal assumes that bradykinin chronically upregulates nitric oxide releasability. Acute release of nitric oxide by activation of endothelial nitric oxide synthase via bradykinin receptor stimulation is well established (15), but the upregulation of EDRF releasability in response to stimuli not acting via the bradykinin receptor [such as acetylcholine and histamine (44, 45)] by chronic bradykinin is not established. Several mechanisms appear possible: bradykinin could stimulate the formation of interleukines (106, 109) and thereby stimulate the synthesis of inducible nitric oxide synthase in the vasculature (6, 60), but upregulated inducible nitric oxide synthase should mainly enhance basal EDRF formation; alternatively, chronic bradykinin exposure could upregulate the constitutive endothelial nitric oxide synthase (such an upregulation has been observed with chronic shear stress in endothelial cells), but this remains to be shown; finally, other chronic endothelial alteration besides altered expression of nitric oxide synthases could affect releasability of EDRF, which may consist of nitric oxide in low molecular nitrosyl iron complexes (15, 110). The exact mechanism of the enhanced EDRF releasability during chronic ACE inhibition in pressure-distended arteries remains unidentified, but it is in agreement with the improvement of EDRF releasability by nonhypotensive chronic ACE inhibition in experimental models of hypercholesterinemia (7, 111).

Thus, our hypothetic scheme predicts two ACE-releated pathways, acting in concert during the trophic adaptation of pressure distended elastic arteries in reactive hypertension: the enhanced formation of the trophic factor angiotensin-II and the reduced releasability of the antitrophic signal nitric oxide due to reduced availability of bradykinin. As a maladaptive corollary of the reduced EDRF availability, the susceptibility of hypertension-exposed, distended arteries to atherogenesis is enhanced. This should not occur in the autoregulating, smaller muscular arteries, which do not uregulate ACE in the coarctation hypertension (see above). Accordingly, our hypothesis assumes that the hypertension-induced upregulation of vascular ACE plays a central role in the localization of atheroma development in presence of circulating risk factors such as hypercholesterolemia (Fig. 1).

Our hypothesis as outlined in Fig. 1 predicts that the endothelial protection by ACE inhibition is bradykinin mediated. We tested this prediction by treating rats with coarctation-induced hypertension for 2 weeks with nonhypotensive angiotensin receptor type-1 blockade (AT_1-blockade), using losartan or ICI-D8731. Much to our surprise, we observed a normalization of the depressed endothelial dilation together with a normalization of the elevated aortic ACE activity (46). This lowering of the elevated aortic ACE activity under chronic AT_1-blockade was not due to some ACE-blocking side effects of the drugs: both antagonists did not alter tissue ACE activity, when given acutely for 1 day. Obviously, the chronic AT_1-blockade caused a downregulation of ACE expression. The mechanism of this downregulation is unknown. We speculate that angiotensin-II could exert a positive impact on aortic ACE expression via activation of AT_1-receptors, probably indirectly by its influence on the subendothelial vascular wall. Chronic AT_1-blockade could attenuate this impact. Regardless of the mechanism, the downregulation of aortic ACE by chronic AT_1-blockade implies that the unexpected endothelial protection by this blockade also is mediated, at least partially, by the bradykinin-dependent pathway in Fig. 1. It would be very important to learn whether a comparable endothelial protection by AT_1-blockade occurs in human vessels. This could explain the impact of a high renin-sodium profile on myocardial infarction (2) and, probably, on atherosclerosis: circulating renin contributes to local vascular angiotensin-II formation after uptake into the vessel wall (102), and thereby could induce upregulation of ACE and downregulation of EDRF releasability.

Another effect of ACE inhibition with potential relevance for the attenuation of atherosclerosis is the suppression of neointima formation in rats after balloon injury of an artery (85). This suppression by ACE inhibition could not be ascribed to the blood pressure lowering by the ACE inhibition (85) and has been confirmed by several groups with various ACE inhibitors (20, 32, 83, 86). Contrary to the initial expectations, this mechanism did not reduce the rate of restenosis after percutaneous coronary angioplasty in humans, but this does not exclude a relevance of this mechanism for the development of atherosclerosis. Similarly as with the protection of endothelium-mediated dilation by ACE inhibition, the suppression of lesion-induced neointima formation appears to be partially bradykinin dependent and sensitive to inhibition by the bradykinin antagonist HOE 140 (32, 37, 83). On the other hand, the formation could also be partially suppressed by AT_1-blockade, indicating a direct involvement of locally formed angiotensin II in this process (35, 37, 62, 81, 86).

A further, rather preliminary observation with potential relevance for the putative antiatherosclerotic potency of ACE inhibition has been published recently: the addition of enalaprilat or ramipril to cell lines in culture enhanced the transcription of low density lipoprotein receptors and potentiated the transcription and translation of these receptors induced by platelet-derived growth factor β-chain (11). However, it was not clarified whether the two inhibitors acted by suppressing angiotensin-II-formation, by suppressing bradykinin degradation or by an unspecific mechanism in these cell systems. Therefore, the biological relevance of these observations remains completely unclear.

In rabbits with heritable hyperlipidemia or in rabbits or monkeys with diet-induced hypercholesterolemia, chronic treatment by ACE inhibition could reduce the atheroma formation (1, 24, 25, 91), as could be expected from the experimental studies summarized in Table 3. The role of ACE-associated hypotension in this antiatherosclerotic action is not quite clear, and the mechanism operative in the rabbit with extremely high cholesterol levels must not be operative in patients. However, several clinical prospective multicenter studies with chronic ACE inhibition are presently performed, designed to delineate the antiatherosclerotic potential of ACE inhibition in humans. The preliminary experimental experience with endothelial protection by AT_1-blockade suggests that this principle should also be tested for its antiatherosclerotic potency in humans.

Our scheme of hypotheses in Fig. 1, though largely speculative, could help to bring activity of the vascular renin angiotensin system, suppression of endothelial EDRF releasability, susceptibility for atherogenesis and trophic adaptation to hypertension-induced arterial distension into a logical context. Future work will show whether the scheme is appropriate.

Table 3. Putatively antiatherosclerotic mechanisms of renin-angiotensin system blockade.

- Bradykinin-dependent normalization of depressed EDRF releasability ("endothelial protection") by nonhypotensive chronic ACE inhibition in experimental hypertension (58, 59)

- Normalization of depressed EDRF releasability ("endothelial protection") and of elevated vascular ACE expression by nonhypotensive chronic AT_1-blockade in experimental hypertension (99)

- Attenuation of balloon-induced neointima formation by chronic ACE inhibition, partially bradykinin dependent (100–105)

- Attenuation of balloon-induced neointima formation by chronic AT_1-blockade (103, 104, 106–108)

- Enhancement of LDL-receptor expression by ACE inhibitors in cultured cell lines in vitro (109)

- Reduction in atheroma development in heritable or diet-induced extreme hypercholesterolemia by chronic ACE inhibition (110–113)

References

1. Aberg G, Ferrer P (1990) Effects of captopril on atherosclerosis in cynomolgus monkeys. J Cardiovasc Pharmacol 15: S65–S72
2. Alderman MH, Madhavan S, Ooi WL, Cohen H, Sealy JE, Laragh JH (1991) Association of the renin-sodium profile with the risk of myocardial infarction in patients with hypertension. N Engl J Med 324: 1098–1104
3. Assender JW, Southgate KM, Newby AC (1991) Does nitric oxide inhibit smooth muscle proliferation? J Cardiovasc Pharmacol 17 (suppl 3): S104–S107
4. Azuma JA, Ishikawa M, Sekizaki S (1986) Endothelium-dependent inhibition of platelet aggregation. Br J Pharmacol 88: 411–415
5. Bath PM, Hassall DG, Gladwin AM, Palmer RM, Martin JF (1991) Nitric oxide and prostacyclin: divergence of inhibitory effects on monocyte chemotaxis and adhesion to endothelium in vitro. Arterioscler Thromb 11: 254–260
6. Beasley D, Schwartz JH, Brenner BM (1991) Interleukin 1 induces prolonged L-arginine-dependent cyclic guanosine monophosphate and nitrite production in rat vascular smooth muscle cells. J Clin Invest 87: 602–608
7. Becker RHA, Wiemer G, Linz W (1991) Preservation of endothelial function by ramipril in rabbits on a long-term atherogenic diet. J Cardiovasc Pharmacol 18 (suppl 2): S110–S115
8. Beierwaltes WH, Carretero OA (1989) Kinin antagonist reverses converting enzyme inhibitor-stimulated vascular prostaglandin I2 synthesis. Hypertension 13: 754–758
9. Berg T, Carretero OA, Scicli AG, Tilley B, Stewart JM (1989) Role of kinin in regulation of rat submandibular gland blood flow. Hypertension 14: 73–80
10. Berliner JA, Territo MC, Sevanian A, Ramin S, Kim JA, Bamshad B, Esterson M, Fogelman AM (1990) Minimally modified low density lipoprotein stimulates monocyte endothelial interactions. J Clin Invest 85: 1260–1266
11. Block LH, Keul R, Crabos M, Ziesche R, Roth M (1993) Transcriptional activation of low density lipoprotein receptor gene by angiotensin-converting enzyme inhibitors and Ca2+-channel blockers involves protein kinase C isoforms. Proc Nat Acad Sci USA 90: 4097–4101
12. Böhme E, Jung R, Melcher I (1974) Guanylate cyclase in human platelets. Nature 11: 1–2
13. Busse R, Lückhoff A, Bassenge E (1987) Endothelium-derived relaxant factor inhibits platelet activation. Naunyn-Schmiedebergs Arch Pharmacol 336: 566–571
14. Busse R, Lamontagne D (1991) Endothelium-derived bradykinin is responsible for the increase in calcium produced by angiotensin-converting enzyme inhibitors in human endothelial cells. Naunyn Schmiedebergs Arch Pharmacol 344: 126–129
15. Busse R, Mülsch A, Fleming I, Hecker M (1993) Mechanisms of nitric oxide release from the vascular endothelium. Circulation 87 (suppl V): V18–V25
16. Cachofeiro V, Nasjletti A (1991) Increased vascular responsiveness to bradykinin in kidneys of spontaneously hypertensive rats: Effect of Nw-nitro-L-arginine. Hypertension 18: 683–688
17. Cachofeiro V, Sakakibara T, Nasjletti A (1992) Kinins, nitric oxide, and the hypotensive effect of captopril and ramiprilat in hypertension. Hypertension 19: 138–145
18. Calver A, Collier J, Moncada S, Vallance P (1992) Effect of local intraarterial N-G-monomethyl-L-arginine in patients with hypertension: the nitric oxide dilator mechanism appears abnormal. J Hypertens 10: 1025–1031
19. Cambien F, Poirier O, Lecerf L, Evans A, Cambon JP, Arveiler D, Lut G, Bard JM, Bara L, Ricard F, Piret L, Amouyel P, Ahlenc-Gelas F, Soubrier F (1992) Deletion polymorphism in the gene for angiotensin converting enzyme is a potent risk factor for myocardial infarction. Nature 359: 641–644
20. Capron L, Heudes D, Charjara A, Bruneval P (1991) Effect of ramipril, an inhibitor of angiotensin converting enzyme, on the response of rat thoracic aorta to injury with a balloon catheter. J Cardiovasc Pharmacol 18: 207–211
21. Carbonell LF, Carretero OA, Stewart JM, Scicli AG (1988) Effect of a kinin antagonist on the acute antihypertensive activity of enalaprilat in severe hypertension. Hypertension 11: 239–243

22. Carretero OA, Scicli AG (1991) Local hormonal factors (intracrine, paracrine, autocrine) in hypertension. Hypertension 18 (suppl I): I58–I69
23. Chin J, Azhar S, Hoffman BB (1992) Inactivation of endothelial derived relaxing factor by oxidized lipoproteins. J Clin Invest 89: 10–18
24. Chobanian AV, Haudenschild CC, Nickerson C, Drago R (1990) Antiatherogenic effect of captopril in the Watanabe heritable hyperlipidemic rabbit. Hypertension 15: 327–331
25. Chobanian AV, Haudenschild CC, Nickerson C, Hope S (1992) Trandolapril inhibits atherosclerosis in the Watanabe heritable hyperlipidemic rabbit. Hypertension 20: 473–477
26. Clozel M, Kuhn H, Hefti F (1990) Effects of angiotensin converting enzyme inhibitors and of hydralazine on endothelial function in hypertensive rats. Hypertension 16: 532–540
27. Cohn JN (1992) The prevention of heart failure: a new agenda. N Engl J Med 327: 725–727
28. Cooke JP, Singer AH, Tsao P, Zera P, Rowan RA, Billingham ME (1992) Antiatherogenic effects of L-arginine in the hypercholesterolemic rabbit. J Clin Invest 90: 1168–1172
29. Cushing SD, Berliner JA, Valente AJ, Territo MC, Navab M, Parhami F, Gerrity R, Schwarz CJ, Fogelman AM (1990) Minimally modified low density lipoprotein induces monocyte chemotactic protein 1 in human endothelial cells and smooth muscle cells. Proc Natl Acad Sci USA 87: 5134–5138
30. Cybulsky MI, Gimbrone MA (1991) Endothelial expression of a mononuclear leukocyte adhesion molecule during atherogenesis. Science 251: 788–791
31. Danckwardt L, Shimizu I, Bönner G, Rettig R, Unger T (1990) Converting enzyme inhibition in kinin-deficient Brown Norway rats. Hypertension 16: 429–435
32. DeBlois D, Lombardi DM, Garvin MA, Schwartz SM (1992) Inhibition by ramipril of intimal hyperplasia in the denuded rat carotid is reserved by Hoe 140, a kinin B2 receptor antagonist (abstr). Circulation 86 (suppl I): I226
33. De'Mey JG, Gray SD (1985) Endothelium-dependent reactivity in resistance vessels. Prog Appl Microcirc 88: 181–187
34. Dohi Y, Thiel MA, Bühler FR, Lüscher TP (1990) Activation of endothelial L-arginine pathway in resistance arteries: effect of age and hypertension. Hypertension 16: 170–179
35. Dzau VJ, Gibbons GH, Pratt RE (1991) Molecular mechanisms of vascular reninangiotensin system in myointimal hyperplasia. Hypertension 18 (suppl II): II100–II105
36. Erdös EG (1990) Angiotensin I converting enzyme and the changes in our concepts through the years: Lewis K Dahl memorial lecture. Hypertension 16: 363–370
37. Fahrhy RD, Ho KL, Carretero OA, Scicli AG (1991) Kinins mediate the antiproliferative effect of ramipril in rat carotid artery. Biochem Biophys Res Commun 182: 283–288
38. Farhy RD, Carretero OA, Ho KL, Scicli AG (1993) Role of kinins and nitric oxide in the effects of angiotensin converting enzyme inhibitors on neointima formation. Circ Res 72: 1202–1210
39. Furlong B, Henderson AH, Leweis MJ, Smith JA (1987) Endothelium-derived relaxant factor inhibits in vitro platelet aggregation. Br J Pharmacol 90: 687–692
40. Galle J, Bassenge E (1991) Effects of native and oxidized low-density lipoproteins on endothelium-dependent and endothelium-independent vasomotion. Basic Res Cardiol 86 (suppl): 127–142
41. Garg UC, Hassid A (1989) Nitric oxide-generating vasodilators and 8-bromo-cyclic guanosine monoposphate inhibit mitogenesis and proliferation of cultured rat vascular smooth muscle cells. J Clin Invest 83: 1774–1777
42. Geritty RG (1981) The role of the monocyte in atherogenesis. I: Transition of bloodborne monocytes into foam cells in fatty lesions. Am J Pathol 103: 181–190
43. Geritty RG (1981) The role of the monocyte atherogenesis. II: Migration of foam cells from atherosclerotic lesions. Am J Pathol 103: 191–200
44. Goetz RM, Krivokuca M, Holtz J (1992) Local activity of angiotensin converting enzyme and local endothelium-dependent dilatory reactivity in coarctation hypertension (abstr). Circulation 86: I558
45. Goetz RM, Studer R, Reinecke H, Holtz J (1993) Enhanced gene expression of angiotensin converting enzyme accounts for endothelial dilatory dysfunction in aorta of rats with coarctation hypertension. Eur Heart J 14 (suppl) 347

46. Goetz RM, Kuch M, Holtz J (1993) Chronische nicht-hypotensive Angiotensin-Rezeptor-Blockade in Hochdruck Ratten: Normalisierung der endothelialen Dilatation (abstr) Z Kardiol 82 (suppl 1): 133

47. Gryglewski RJ, Palmer RMJ, Moncada S (1986) Superoxide anion is involved in the breakdown of endothelium-derived vascular relaxing factor. Nature 320: 454–456

48. Guerra R, Brotherton AFA, Goodwin PJ, Clark CR, Armstrong JL, Harrison DG (1989) Mechanisms of abnormal endothelium-dependent vascular relaxation in atherosclerosis: implications for altered autocrine and paracrine functions of EDRF. Blood Vessels 26: 300–314

49. Hajj-ali AF, Zimmerman BG (1992) Nitric oxide participation in renal hemodynamic effect of angiotensin converting enzyme inhibitor lisinopril. Eur J Pharmacol 212: 279–281

50. Hawkins DJ, Meyrick BO, Murray JJ (1988) Activation of guanylate cyclase and inhibition of platelet aggregation by endothelium-derived relaxing factor released from cultured cells. Biochim Biophys Acta 969: 289–296

51. Heinecke JW, Baker L, Rosen H, Chait A (1986) Superoxide-mediated modifications of low density lipoprotein by arterial smooth muscle cells. J Clin Invest 77: 757–761

52. Henning B, Chow CK (1988) Lipid peroxidation and endothelial cell injury: implications in atherosclerosis. Free Radical Biol Med 4: 99–106

53. Hirooka Y, Imaizumi T, Masaki H, Ando S, Harada S, Momohara M, Takeshita A (1992) Captopril improves impaired endothelium-dependent vasodilation in hypertensive patients. Hypertension 20: 175–180

54. Hogan JC, Lewis MJ, Henderson AH (1988) In vivo EDRF activity influences platelet function. Br J Pharmacol 94: 1020–1022

55. Holtz J, Goetz RM (1991) Peptides in coronary circulation: basis for therapeutic strategies. Eur Heart J 12 (suppl F): 112–120

56. Itoh H, Pratt RE, Dzau VJ (1990) Atrial atriuretic polypeptide inhibits hypertrophy of vascular smooth muscle cells. J Clin Invest 86: 1690–1697

57. Itoh H, Pratt RE, Dzau VJ (1991) Interaction of atrial natriuretic polypeptide and angiotensin II on protooncogenene expression and vascular cell growth. Biochem Biophys Res Commun 176: 1601–1609

58. Johnson A, Lermioglu F, Garg UC, Morgan-Boyd R, Hassid A (1988) A novel biological effect of atrial natriuretic hormone: Inhibition of mesangial cell proliferation. Biochem Biophys Res Commun 152: 993–997

59. Johnson G, Taso PS, Mulloy D, Lefer AM (1990) Cardioprotective effects of acidified sodium nitrite in myocardial ischemia with reperfusion. J Pharmacol Exp Ther 252: 35–41

60. Joly GA, Schini VB, Vanhoutte PM (1992) Balloon injury and interleukin-1β induce nitric oxide synthase activity in rat carotid arteries. Circ Res 71: 331–338

61. Kubes P, Suzuki M, Granger DN (1991) Nitric oxide: an endogenous modulator of leukocyte adhesion. Proc Natl Acad Sci USA 88: 4651–4655

62. Laporte S, Gervais A, Escher E (1991) Angiotensin II antagonists prevent the myoproliferative response after vascular injury (abstr). FASEB J 5: A869

63. Lehr HA, Hübner C, Finckh B, Angermüller S, Nolte D, Beisiegel U, Kohlschütter A, Messmer K (1991) Role of leukotrienes in leukocyte adhesion following systemic administration of oxidatively modified human low density lipoprotein in hamsters. J Clin Invest 88: 9–14

64. Lehr HA, Becker M, Marklund SL, Hübner C, Arfors E, Kohlschütter A, Messmer K (1992) Superoxide-dependent stimulation of leukocyte adhesion by oxidatively modified LDL in vivo. Arteriosclerosis 12: 824–829

65. Linder L, Kiowski W, Bühler FR, Lüscher TF (1990) Indirect evidence for release of endothelium-derived relaxing factor in the human forearm circulation in vivo: blunted response in essential hypertension. Circulation 81: 1762–1767

66. Linz W, Schölkens BA (1990) A specific B2-bradykinin receptor antagonist HOE 140 abolishes the antihypertrophic effect of ramipril. Br J Pharmacol 105: 771–772

67. Lockette W, Otsuka Y, Carretero O (1986) The loss of endothelium-dependent vascular relaxation in hypertension. Hypertension 8 (suppl II): II61–II66

68. Lüscher TF, Vanhoutte PM (1986) Endothelium-dependent responses to platelets and serotonin in spontaneously hypertensive rats. Hypertension 8 (suppl II): II55–II60
69. Lüscher TF, Boulanger CM, Yang Z, Noll G, Dohi Y (1993) Interactions between endothelium-derived relaxing and contracting factors in health and cardiovascular disease. Circulation 87 (suppl V): V36–V44
70. Lüscher TF, Haefeli WE (1993) L-Arginine in the clinical arena: Tool or remedy? Circulation 87: 1746–1748
71. Mao S, Yates M, Lambert L, Whitten J, McDonald I, Ku G, Mano M (1992) Nitric oxide protects the oxidative modification of low density lipoprotein by macrophages. FASEB J 6: A549
72. Matsubara T, Ziff M (1986) Increased superoxide anion release from human endothelial cells in response to cytokines. J Immunol 137: 3295–3298
73. Matsuda Y, Hirata K, Inoue N, Suematsu M, Kawashima S, Akita H, Yokoyama M (1993) High density lipoprotein reverses inhibitory effect of oxidized low density lipoprotein on endothelium-dependent arterial relaxation. Circ Res 72: 1103–1109
74. Miller VM (1990) Does antihypertensive therapy improve the function of the vascular endothelium? Hypertension 16: 541–543
75. Minor RL, Myers PR, Guerra R, Bates JN, Harrison DG (1990) Diet-induced atherosclerosis increases the release of nitrogen oxides from rabbit aorta. J Clin Invest 86: 2109–2116
76. Miyazaki M, Okunishi H, Okamura T, Toda N (1987) Elevated vascular angiotensin converting enzyme in chronic two-kidney, one clip hypertension in the dog. J Hypertens 5: 155–160
77. Miyazaki M, Okamura T, Toda N (1988) Role of vascular angiotensin converting enzyme in hypertension. J Hypertens 6 (suppl 3): S13–S15
78. O'Connor KJ, Knowles RG, Patel KD (1991) Nitrovasodilators have proliferative as well as antiproliferative effects. J Cardiovasc Pharmacol 17 (suppl 3): S100–S103
79. Okamura T, Miyazaki M, Inagami T, Toda N (1986) Vascular renin-angiotensin system in two-kidney, one clip hypertensive rats. Hypertension 8: 560–565
80. Okunishi H, Kawamoto T, Kurobe Y, Oka Y, Ishii K, Tanaka T, Miyazaki M (1991) Pathogenetic role of vascular angiotensin converting enzyme in the spontaneously hypertensive rat. Clin Exp Pharmacol Physiol 18: 649–659
81. Osterrieder W, Müller RKM, Powell JS, Clozel JP, Hefti F, Baumgartner HR (1991) Role of angiotensin II in injury-induced neointima formation in rats. Hypertension 18 (suppl II): II60–II64
82. Panza JA, Quyyumi AA, Brush JE, Epstein SE (1990) Abnormal endothelium-dependent vascular relaxation in patients with essential hypertension. N Engl J Med 323: 22–27
83. Panza JA, Casino PR, Kilcoyne CM, Quyyumi AA (1993) Role of endothelium-derived nitric oxide in the abnormal endothelium-dependent vascular relaxation of patients with essential hypertension. Circulation 87: 1468–1474
84. Pfeffer MA, Braunwald E, Moye LA, Basta L, Brown EJ, Cuddy TE, Davis BR, Galtman EM, Goldman S, Flaker GC, Klein M, Lamas GA, Packer M, Rouleau J, Rouleau JL, Rutherford J, Wertheimer JH, Hawkins CM (1992) Effect of captopril on mortality and morbidity in patients with left ventricular dysfunction after myocardial infarction: Results of the survival and ventricular enlargement trial. N Engl J Med 327: 669–677
85. Powel JS, Clozel JP, Müller RK, Kuhn H, Hefti F, Hosang M, Baumgartner HR (1989) Inhibitors of angiotensin-converting enzyme prevent myointimal proliferation after vascular injury. Science 245: 186–188
86. Prescott MF, Webb RL, Reidy MA (1991) Angiotensin-converting enzyme inhibitor versus angiotensin II, AT1 receptor antagonist: Effects on smooth muscle cell migration and proliferation after balloon catheter injury. Am J Pathol 139: 1291–1296
87. Quilley J, Duchin KL, Hudes EM, McGiff JC (1987) The antihypertensive effect of captopril in essential hypertension: Relationship to prostaglandins and the kallikrein-kinin system. J Hypertens 5: 121–128

88. Radomski MW, Palmer RMJ, Moncada S (1987) Comparative pharmacology of endothelium-derived relaxing factor, nitric oxide, and prostacyclin in platelets. Br J Pharmacol 92: 181–187

89. Ross R (1986) The pathogenesis of atherosclerosis. N Engl J Med 314: 488–500

90. Schray-Utz B, Zeiher A, Busse R (1993) The expression of monocyte chemoattractant protein (MCP-1) mRNA in human endothelial cells in modulated by nitric oxide. FASEB J 7: A130

91. Schuh JR, Blehm DJ, Friedrich GE, McMahon EG, Blaine EH (1992) Differential effects of renin angiotensin system blockade on atherogenesis in cholesterol-fed rabbits (abstr). FASEB J 6: A1031

92. Scott-Burden T, Hahn AWA, Resink TJ, Bühler FR (1990) Modulation of extracellular matrix by angiotensin II: stimulation of glycoconjugate synthesis and growth in vascular smooth muscle cells. J Cardiovasc Pharmacol 16: S36–S41

93. Scott-Burden T, Schini VB, Vanhutte PM (1992) Induction by interleukin-1β of nitric oxide synthase leads to inhibition of mitogen-stimulated DNA synthesis in cultured vascular smooth muscle cells. In: Biology of nitric oxide, Moncada S, Marletta MA, Hibbs JB (eds) Portland Press, London, pp 48–52

94. Scott-Burden T, Vanhoutte PM (1993) The endothelium as a regulator of vascular smooth muscle proliferation. Circulation 87 (suppl V): V51–V55

95. Sealey JE, Blumenfeld JD, Bell GM, Pecker MS, Sommers SC, Laragh JH (1988) On the renal basis for essential hypertension: nephron heterogeneity with discordant renin secretion and sodium excretion causing a hypertensive vasoconstriction-volume relationship. J Hypertens 6: 763–777

96. Shiota N, Miyazaki M, Okunishi H (1992) Increase of angiotensin converting enzyme gene expression in the hypertensive aorta. Hypertension 20: 168–174

97. Shultz PJ, Raij L (1989) Effects of antihypertensive agents on endothelium-dependent and endothelium-independent relaxations. Br J Clin Pharmacol 28: 1515–1525

98. Simpson PJ, Todd RF, Fantone JC, Mickelson JK, Griffin JD, Lucchesi BR (1988) Reduction of experimental canine myocardial reperfusion injury by a monoclonal antibody (anti-Mol, anti Cd11b) that inhibits leukocyte adhesion. J Clin Invest 81: 624–629

99. Simpson PJ, Todd RF, Mickelson JK, Fantone JC, Gallagher KP, Lee KA, Tamura Y, Cronin M, Lucchesi BR (1990) Sustained limitation of myocardial reperfusion injury by a monoclonal antibody that alters leukocyte function. Circulation 81: 226–237

100. SOLVD Investigators (1991) Effect of enalapril on survival in patients with reduced left ventricular ejections fractions and congestive heart failure. N Engl J Med 325: 293–302

101. Steinberg D (1983) Lipoproteins and atherosclerosis: a look back and a look ahead. Arteriosclerosis 3: 283–301

102. Taddei S, Virdis A, Abdel-Haq B, Giovannetti R, Duranti P, Arena AM, Favilla S, Salvetti A (1993) Indirect evidence for vascular uptake of circulating renin in hypertensive patients. Hypertension 21: 852–860

103. Tagawa H, Tomoike H, Nakamura M (1991) Putative mechanisms of the impairment of endothelium-dependent relaxation of the aorta with atheromatous plaque in heritable hyperlipidemic rabbits. Circ Res 68: 330–337

104. Tanner FC, Noll G, Boulanger CM, Lüscher TF (1991) Oxidized low density lipoproteins inhibit relaxations of porcine coronary arteries: role of scavenger receptor and endothelium-derived nitric oxide. Circulation 83: 2012–2020

105. Tesfariam B, Halpern W (1988) Endothelium-dependent and endothelium-independent vasodilation in resistance arteries from hypertensive rats. Hypertension 11: 440–444

106. Tiffany CW, Burch RM (1989) Bradykinin stimulates tumor necrosis factor and interleukin-1 release from macrophages. FEBS Lett 247: 189–192

107. Tomita T, Ezika M, Miwa M, Nakamura K, Inoue Y (1990) Rapid and reversible inhibition by low density lipoprotein of the endothelium-dependent relaxation to hemostatic substances in procine coronary arteries. Circ Res 58: 552–564

108. Vallance P, Leone A, Calver A, Collier J, Moncada S (1992) Accumulation of an endogenous inhibitor of nitric oxide synthesis in chronic renal failure. Lancet 339: 573–575

109. Vandekerckhove F, Opdenakker G, Van'Ranst M, Lenaerts JP, Put W, Billiau A, Van'Damme J (1991) Bradykinin induces interleukin-6 and synergizes with interleukin-1. Lymphokine Cytokine Res 10: 285–289
110. Vanin AF (1991) Endothelium-derived relaxing factor is a nitrosyl iron complex with thiol ligands. FEBS Lett 289: 1–3
111. Webb RC, Finta KM, Fisher M, Lee L, Pitt B (1992) Ramipril reverses impaired endothelium-dependent relaxation in arteries from rabbits fed an atherogenic diet (abstr). FASEB J 6: 3022
112. Wiemer G, Schölkens BA, Becker RHA, Busse R (1991) Ramiprilat enhances endothelial autacoid formation by inhibiting breakdown of endothelium-derived bradykinin. Hypertension 18: 558–563
113. Winquist RJ, Bunting PB, Baskin EP, Wallace AA (1984) Decreased endothelium-dependent relaxations in New Zealand genetic hypertensive rats. J Hypertens 2: 541–545

Authors' address:
Prof. Dr. J. Holtz
Institut für Pathophysiologie
Martin-Luther-Universität Halle-Wittenberg
Magdeburger Str. 6
D-06112 Halle-Saale
FRG

Endothelial dysfunction in atherosclerosis

D. G. Harrison

Department of Internal Medicine, Emory University School of Medicine and Atlanta Veterans Administration Hospital

Summary: Endothelial regulation of vasomotor tone occurs largely via the release of nitric oxide or a closely related compound. This process is strikingly altered in a variety of disease states, and alterations of vasomotion may be responsible for the development of hypertension, altered tissue perfusion, and an enhanced propensity for vasoconstriction in several common disorders. In hypercholesterolemia and atherosclerosis, this alteration of vasomotor control occurs not only in larger vessels, but in the microcirculation. Explanations for impaired endothelium-dependent vascular relaxations in hypercholesterolemia include impairments in endothelial cell signal transduction, deficiencies in the substrate (arginine) for the enzyme nitric oxide synthase, alterations in the nitric oxide synthase enzyme or one of its co-factors, and excess destruction of nitric oxide by the superoxide anion. In this review, evidence for these alterations will be considered, potential interventions for restoring endothelium-dependent relaxations examined, and the possible impact of endothelial dysfunction in atherosclerosis considered.

Key words: Hypercholesterolemia – vasodilatation – nitric oxide – superoxide anion

Introduction

In 1980, Furchgott and Zawadski first showed that the endothelium must be intact for acetylcholine to produce vascular relaxation (17). Subsequently, numerous neurohumoral and pharmacological agents were shown to mediate vasodilation via the endothelium (5, 6, 9, 10, 51, 56). It is now recognized that the endothelium releases a potent labile nonprostanoid vasodilating agent in response to numerous vasoactive stimuli that either causes vasodilatation or modulates vasoconstriction (18, 19). This factor, subsequently termed the endothelium-derived relaxing factor, recently has been shown to be either nitric oxide (39) or a compound with a nitric oxide moiety within its structure (37). NO serves to activate vascular smooth muscle soluble guanylate cyclase, increasing cGMP and decreasing intracellular Ca^{++}, among other effects. There may be multiple EDRFs (15, 35, 44), and the endothelium also releases a substance that causes vascular smooth muscle hyperpolarization (13).

Supported by NIH grants HL32717 and HL39006, and a Merit Grant from the Veterans Administration. Dr. Harrison is an Established Investigator of the American Heart Association.

A fascinating aspect of the biology of the endothelium-derived nitric oxide ED(NO) is that its regulatory influences are abnormal in a number of disease states. Probably the most thoroughly studied of these are atherosclerosis and hypercholesterolemia.

In this review, abnormalities of endothelial regulation of vasomotor tone due to atherosclerosis and hypercholesterolemia will be reviewed. Recent observations regarding the production of nitrogen oxides from normal and atherosclerotic vessels will be presented and studies of endothelial function in the coronary microcirculation will be discussed. Finally, we will speculate as to the mechanisms underlying abnormal endothelium-dependent vascular relaxation in atherosclerosis.

Impaired endothelium-dependent relaxations in atherosclerosis

Animal Studies

Large Conduit Arteries and *chronic exposure to high cholesterol:* In 1986 and 1987, several groups demonstrated that endothelium-dependent vascular relaxations are impaired in animals with experimentally induced atherosclerosis. Habib and co-workers studied aortic rings from cholesterol-fed rabbits and found that relaxations to acetylcholine were markedly abnormal (24). Verbuerren and co-workers studied rabbits fed cholesterol for 8 and 16 weeks (54). These investigators found that relaxations to acetylcholine were dramatically impaired in both groups while relaxations to nitroglycerin were slightly impaired only in the longer term group. Contractions to serotonin were selectively enhanced in cholesterol-fed animals. Jayakody et al. also showed that cholesterol feeding of rabbits decreased aortic relaxations to acetylcholine (28). Studies by Freiman et al. (16) and Harrison et al. (23), showed that endothelium-dependent vascular relaxations to acetylcholine, thrombin, and the calcium ionophore A23187 were impaired in iliac arteries from monkeys with diet-induced atherosclerosis (Fig. 1), while relaxations to nitroglycerin were normal.

In most of these studies, histologic examination of the endothelium in the experimental groups revealed either no abnormality or only subtle changes in the endothelium. Specifically, endothelial cell denudation was minimal or absent. Thus, it was concluded from these studies that hypercholesterolemia and diet-induced atherosclerosis produces a functional rather than anatomic abnormality of the endothelium.

Effect of Hypercholesterolemia on the Coronary Microcirculation

All of the early work relevant to endothelial vascular function involved large conduit vessels (aorta, iliac arteries, proximal coronary arteries) which are not generally involved in the direct regulation of tissue perfusion. The resistance vasculature does not develop overt atherosclerosis. Thus, it seemed possible that endothelium-dependent vascular relaxation may be normal in this segment of the circulation despite abnormalities in larger vessels in the setting of chronic hyperlipidemia and atherosclerosis.

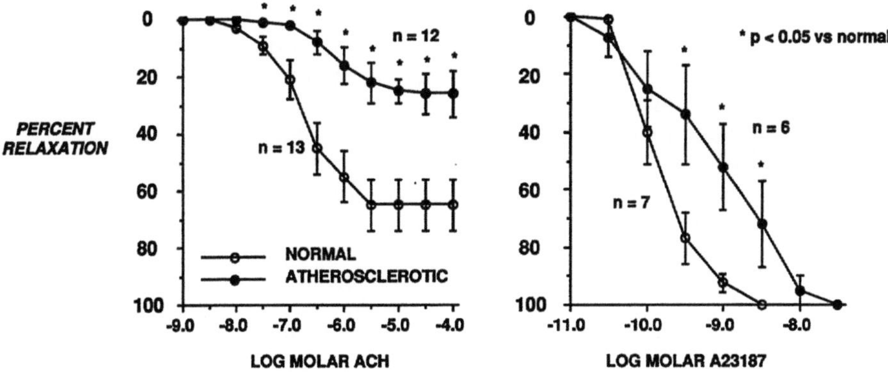

Fig. 1. Responses of normal and atherosclerotic iliac arteries from Cynomolgus monkeys to acetylcholine (left panel) and the calcium ionophore A23187 (right panel). Atherosclerosis was produced by feeding a 0.7 % cholesterol diet for 18 months. Vessels were studied as isolated rings in organ chambers and preconstricted with prostaglandin F2α. Vasorelaxing agents were added in increasing concentrations. (Reproduced with permission from the American Heart Association and the authors.)

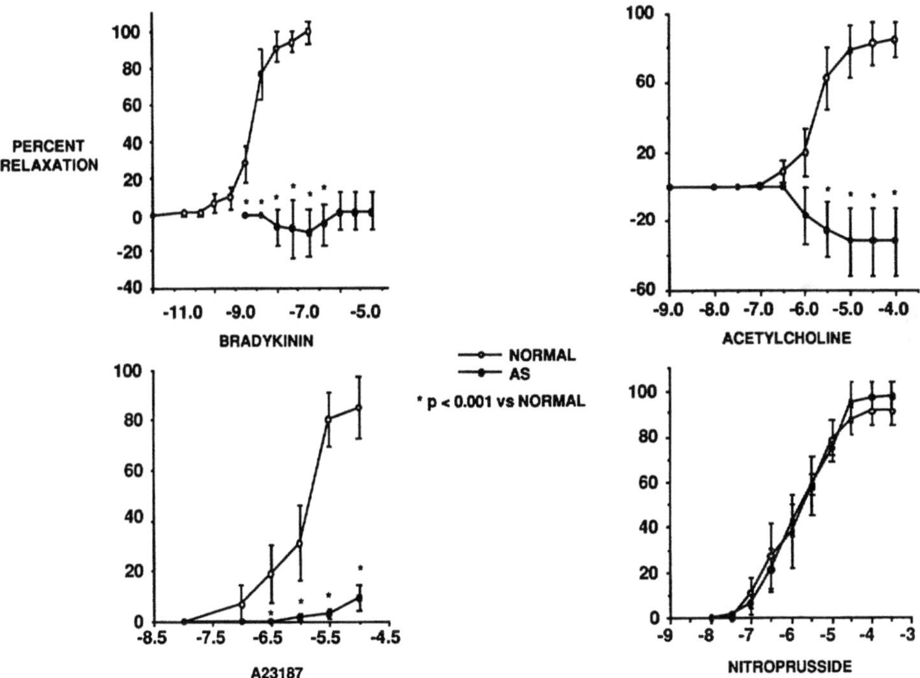

Fig. 2. Responses of coronary microvessels from Cynomolgus monkeys. Chronic hypercholesterolemia was produced by feeding the animals a 0.7 % high cholesterol diet for 18 months. Vessels 100 to 200 μm in diameter were studied in a pressurized state following preconstriction with the thromboxane mimetic U46619. (Reproduced with permission from the American Heart Association and the authors.)

Several studies have now demonstrated that hypercholesterolemia can markedly alter endothelial function in the microcirculation. Yamamoto showed that the relaxations to acetylcholine were strikingly abnormal in 25 μm diameter arterioles of the cremaster muscle of cholesterol-fed rabbits (57). Sellke et al. have studied resistance arteries (100–200 μm in diameter) of cholesterol-fed primates using a unique in vitro microvessel imaging apparatus (46). Prior in vivo studies had shown that these vessels were true coronary resistance arteries (5a). Relaxations to the endothelium-independent vasodilators nitroprusside and adenosine were not altered in these vessels. In contrast, endothelium-dependent vascular relaxations to bradykinin, acetylcholine, and the calcium ionophore A23187 were strikingly abnormal in resistance vessels from cholesterol-fed animals (Fig. 2). Quillen et al. showed that several substances released from aggregating platelets and thrombi produce abnormal responses in coronary microvessels of cholesterol-fed primates (40a). In this study, longer term exposure to hypercholesterolemia resulted in thrombin producing vasoconstriction rather than vasodilatation (Fig. 3). More recently, Kuo et al. sowed strikingly abnormal endothelium-dependent responses to bradykinin and flow in coronary microvessels from pigs with diet-induced hypercholesterolemia (31a). (This study is described more completely below.) These studies of the microcirculation clearly show that hypercholesterolemia may affect endothelium-dependent vascular relaxation in vessels that do not develop overt atherosclerosis. Further, these findings likely have important implications regarding neurohumoral regulation of tissue perfusion in atherosclerosis.

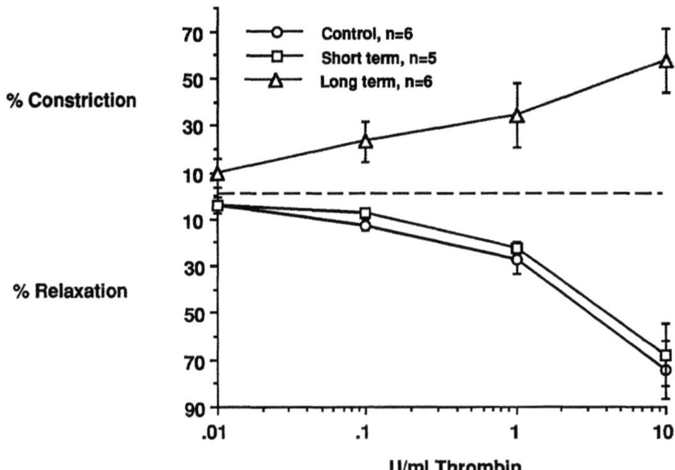

Fig. 3. Effect of long-and short-term hypercholesterolemia on responses of coronary microvessels to thrombin. Hypercholesterolemia was produced by feeding Cynomolgus monkeys a 0.7 % high cholesterol diet for either 8–11 weeks (short-term group) or 18 months (long-term group). Vessels 100 to 200 μm in diameter were studied in a pressurized state following preconstriction with the thromboxane mimetic U46619. (Reproduced with permission from the American Heart Association and the authors.)

90

Studies in humans

In vitro studies

Shortly after the initial observations of impaired endothelium-dependent relaxations in cholesterol-fed animals came evidence that this function of the endothelium is alo impaired in humans with atherosclerosis. Two studies examined endothelium-dependent vascular relaxations of human coronary arteries in vitro. Bossaller et al. studied nonatherosclerotic and atherosclerotic ring segments of coronary arteries from patients undergoing cardiac transplantation (3). Relaxations of atherosclerotic segments to acetylcholine, substance P, and histamine were impaired while relaxations to the calcium ionophore A23187 were preserved. They further found that cyclic GMP accumulation in atherosclerotic segments was impaired to acetylcholine but unaltered in response to calcium ionophore. These investigators proposed that atherosclerosis altered endothelial cell receptor function and that the responses to the calcium ionophore were intact because this agonist bypasses receptor-mediated mechanisms.

Forstermann and co-workers studied endothelium-dependent vascular relaxation in strips of coronary arteries removed from patients undergoing heart transplant (14). These investigators found that responses to all endothelium-dependent vasodilators including substance P, bradykinin, and the calcium ionophore A23187 were abnormal in atherosclerotic segments compared to nonatherosclerotic segments. The reason for the discrepancy between these two studies regarding the effect of the calcium ionophore A23187 remains unclear.

Studies in the Catheterization Laboratory

Several groups have examined the effect of acetylcholine on human coronary arteries at the time of cardiac catheterization. Ludmer et al. showed that acetylcholine, when infused into the left anterior descending coronary artery produced either no response or minimal dilatation of normal coronary vessels, while causing constriction (often severe, near total occlusions) of the atherosclerotic coronary arteries (33). There has been substantial controversy regarding this finding. Okamura and co-workers have reported that somewhat larger concentrations of acetylcholine uniformly caused vasoconstriction (38). This apparent paradox may be explained by the possibility that the group studied by Okamura et al. may have had minimal, nonobstructive atherosclerosis which caused endothelial cell dysfunction. Indeed, acetylcholine was more likely to produce vasoconstriction of an angiographically normal left anterior descending artery if there was angiographically apparent disease of either of the other two major coronary arteries, suggesting that the left anterior descending likely had undetected atherosclerosis (33). Vita et al. have shown that the propensity for vasodilation versus vasoconstriction produced by intracoronary acetylcholine bears a strong relationship to the individual's number of risk factors for atherosclerosis (55).

Probably one of the most potent stimuli for the release of ED(NO) is mechanical shear stress (26, 27, 43). This phenomenon may help maintain shear stress at relatively constant levels during large changes in blood flow. This regulatory function of the endothelium is also abnormal in atherosclerosis. In two independent studies (8, 11), adenosine was infused into the left anterior descending coronary artery of patients at the time of cardiac catheterization. The diameter of the left anterior descending artery proximal to the site of adenosine infusion was measured using quantitative techniques. Among the patients with entirely normal (smooth) coronary arteries, the proximal left anterior descending artery dilated. Since this portion of the vessel was not exposed to adenosine, the observed vasodilatation was assumed to be due to flowmediated release of ED(NO). In contrast, among patients with minimal atherosclerosis, essentially no flow-mediated vasodilation was observed. This finding of impaired flow-mediated vascular relaxation in atherosclerosis has recently been confirmed in monkeys with diet-induced atherosclerosis (34).

Based on these in vitro and in vivo studies of human coronary arteries, it is now apparent that atherosclerosis not only alters endothelium-dependent vascular relaxation in experimental animals with diet-induced atherosclerosis, but also results in a similar defect of endothelial function in the coronary circulation of humans with atherosclerosis. These observations may have important implications regarding the pathogenesis of vascular spasm and alterations of neurohumoral regulation of the circulation in atherosclerosis.

Can alterations of endothelium-dependent relaxation be prevented or corrected in atherosclerotic vessels?

If alterations of endothelium-dependent responses are important in modulating vasomotor tone in atherosclerosis, correction of this abnormality may have important clinical implications. Several studies have addressed this issue. Habib and co-workers examined endothelium-dependent responses to acetylcholine in three groups of rabbits: a control group, a cholesterol-fed group, and a cholesterol-fed group treated with the calcium antagonist PN200110 (24). Cholesterol feeding markedly elevated both plasma cholesterol levels and the level of cholesterol in the rabbit aorta and impaired endothelium-dependent relaxation to acetylcholine. The plasma cholesterol level of the animals treated with a calcium antagonist were elevated to the same degree as the cholesterol-fed group; however, there was a substantial reduction in the amount of cholesterol within the aortic wall. Endothelium-dependent relaxation to acetylcholine was reduced in this group, but not nearly as severely as in the group fed cholesterol alone. Thus it would appear that treatment with calcium antagonist may slow the development of atherosclerosis and partially preserved endothelium-dependent responses.

We have examined the effect of dietary treatment on endothelium-dependent responses to atherosclerosis (22). Three groups of animals were examined in our study. One was a group of Cynomolgus monkeys fed an atherogenic diet for 18 months. A second group of monkeys received an atherogenic diet for 18 months followed by a normal diet for the ensuing 18 months. A third group of monkeys served as control. Within the cholesterol-fed group, there was marked intimal thickening with lipid-laden foam cells and inflammatory cells present in the intima. After regression of atherosclerosis, the intimal area decreased by about 50 % but remained markedly thickened compared to control vessels. In contrast to this modest improvement in intimal thickening, endothelium-dependent responses to both acetylcholine and thrombin were restored entirely to normal (Fig. 4). Responses to nitroglycerin were similar among all groups.

Interestingly, in this study, the regression of intimal thickening was far from complete and yet the abnormality of endothelium-dependent relaxation was eliminated by atherosclerosis regression. Morphological evidence of atherosclerosis regression occurs predominantly when lesions contain large quantities of foam cells and intracellular lipids. When atherosclerotic vessels contain predominantly fibrotic plaque without a large amount of intracellular lipid and foam cells, decreases in lesion size are less likely to occur despite improvement or even correction of the serum lipids. In contrast to the concept that anatomic regression may be limited by the presence of fibrosis in the intima of atherosclerotic vessels, the results from our study of regression suggests that dietary treatment may produce an improvement in functional abnormalities of the vessel in atherosclerosis. This "functional regression" may occur despite only modest decreases in intimal thickness or minimal changes in the gross appearance of the lesion. It is interesting to speculate whether the dietary changes or alterations in medical therapy might promote absorption of lipid from the vessel wall, restore endothelial modulation of vascular smooth muscle reactivity toward normal, and result in resolution of abnormal vasomotor phenomenon in the clinical setting.

Shimokawa and Vanhoutte have shown that dietary cod liver oil improves endothelium-dependent responses in a pig model of focal coronary atherosclerosis produced by a combination of balloon denudation and cholesterol feeding (48). Vekshtein and co-workers have recently presented preliminary data showing that 6

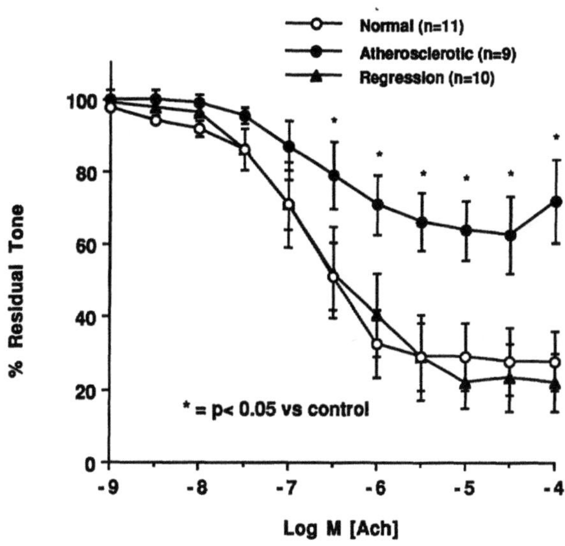

Fig. 4. Effect of cholesterol lowering on endothelium-dependent vascular relaxation to acetylcholine. Cynomolgus monkeys were made atherosclerotic by feeding a high cholesterol diet for 18 months. In some animals regression of atherosclerosis was accomplished by subsequently feeding a normal diet for 18 months. Iliac arteries from each group, and from normal animals, were studied as isolated rings in organ chambers and preconstricted with prostaglandin F2α. Vasorelaxing agents were added in increasing concentrations. (Reproduced with permission from the Journal of Clinical Investigation and the authors.)

mounths of therapy with omega-3 fatty acids can reverse the abnormal coronary constriction to acetylcholine in patients with early atherosclerosis (53).

Several other studies have been performed to specifically correct potential alterations of endothelial cell metabolism which may exist in hypercholesterolemia. These will be discussed in the following paragraphs.

Mechanism of impaired endothelium-dependent relaxation in atherosclerotic blood vessels

Release and Action of ED(NO)

Studies such as those described in the preceding sections, in which intact vessels, intact animals, or the coronary circulation of humans is studied during cardiac catheterization, are not capable of determining if abnormal endothelium-dependent vascular relaxation is due to endothelial dysfunction or if the vascular smooth muscle of atherosclerotic vessels is less responsive to EDRF. To address this question, Guerra et al. (26) employed a bioassay preparation to examine the amount and effect of EDRF released from normal and atherosclerotic rabbit aorta (6 months of cholesterol feeding). EDRF released from atherosclerotic rabbit aorta in response to both acetylcholine and the calcium ionophore A23187 produced only about one-half as much relaxation of denuded normal detector vessel ring segments as in response to EDRF from normal rabbit aorta (Fig. 5a). In contrast, atherosclerotic vessels were found to be supersensitive to EDRF and equally sensitive to nitric oxide compared to normal vessels (Fig. 5b). In addition, the autocrine/paracrine function of EDRF in stimulating the production of cyclic GMP within endothelial cells was also impaired by atherosclerosis. Thus, these bioassay studies clearly showed that the predominant mechanism underlying abnormal endothelium-dependent vascular relaxation is related to either decreased release of EDRF or release of a defective EDRF rather than insensitivity of atherosclerotic smooth muscle to EDRF.

Miscellaneous Mechanisms

It is conceivable that other mechanisms may contribute to abnormal endothelium-dependent relaxation in atherosclerosis. Bioassay studies can only examine the luminal release of EDRF. Since the luminal and abluminal release of EDRF may be quantitatively (2) and qualitatively different (44), it is conceivable that the abluminal diffusion of EDRF may be impaired by the often marked intimal thickening that occurs in the advanced stages of atherosclerosis. It is also possible that EDRF may be degraded by oxygen radicals released from subintimal macrophages and other inflammatory cells present in the atherosclerotic lesion (19, 40, 45). Shimokawa and co-workers have shown that intimal thickening developing as a result of vascular trauma may in fact impair endothelium-dependent relaxations to serotonin, but not to bradykinin or the calcium ionophore A23187 (47).

It is now apparent that the endothelium releases a variety of constrictor factors including several cyclooxygenase prostanoids (36, 49, 50), the superoxide anion (30),

94

angiotensin II (12), and the polypeptide endothelin (58). If these compounds were released in excess in the setting of atherosclerosis, it is possible that they may counteract the concomitant release of EDRF. Studies of rabbit aorta studied both as intact rings and in bioassay have failed to show any evidence for this phenomenon (21).

Potential role of intimal inflammatory cells

The atherosclerotic plaque contains a number of abnormal cellular elements, including leukocytes, macrophages, and adherent platelets, which may release potent vasoactive agents. These include thromboxane A_2 from platelets and leukocytes, leukotrienes from leukocytes, oxygen-derived radicals from leukocytes and macrophages, and vasoconstrictor prostanoids from leukocytes (42, 52). Lopez and co-workers examined the effect of the peptide fmet-leu-phe (fMLP) when infused into the hindlimb of normal and atherosclerotic Cynomolgus monkeys (32). This peptide activates polymorphonucleocytes and macrophage-monocytes. fMLP did not change large artery resistance in normal monkeys, however, it produced pronounced constriction of large arteries in atherosclerotic monkeys. These results suggest that resident inflammatory cells, present in atherosclerotic vessels but not in normal vessels, may contribute to vascular spasm. Furthermore, Lopez et al. found that prostaglandin E_2, which may be released from leukocytes, produced marked constriction of

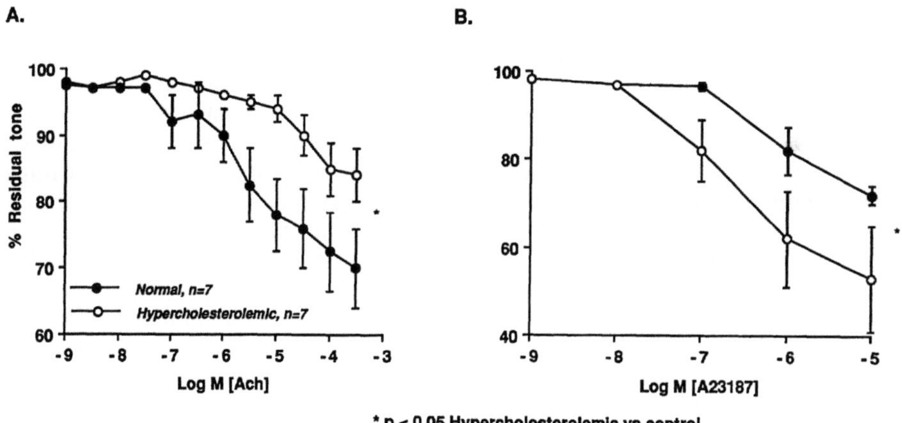

Fig. 5. A) Bioassay experiments examining the release of EDRF from normal and cholesterol-fed rabbit aorta. A) Ring segment of normal rabbit aorta, denuded of endothelium was used as a detector vessel. B) Bioassay experiments examining the sensitivity of normal and cholesterol-fed rabbit aortas of EDRF from a normal segment of rabbit aorta. In the latter experiments the detector segments were perfused with effluent from perfused segments of normal rabbit aorta. The detector segments had been denuded of endothelium, and EDRF release stimulated from the donor segment by administering increasing concentrations of A23187. (Reproduced with permission from Blood Vessels and the authors.)

large arteries in atherosclerotic but not in normal monkeys. Thus, not only do atherosclerotic vessels contain a variety of cell types that may release vasoactive agents, but these vessels have increased responsiveness to at least one vasoactive substance released from leukocytes.

Mechanisms causing altered ED(NO) release in hypercholesterolemia

The defect responsible for this defect in endothelial function in hypercholesterolemia has been the matter of substantial interest. Experimental data have suggested that in atherosclerosis there are deficiencies of membrane signalling within the endothelial cell (13a), a deficiency of arginine (the substrate for endothelium-derived nitrogen oxides (7a, 17a, 42a), or increased degradation of the EDRF by the superoxide anion.

Alterations of arginine metabolism

There has been substantial interest in the hypothesis that in the setting of hypercholesterolemia an abnormality of arginine supply or metabolism exists such that there is a deficiency of arginine in the endothelium. Several groups have now reported that the administration of L-arginine improves responses of resistance vessels to endothelium-dependent vasodilators, both in hypercholesterolemic animals and in humans (11a, 17a, 31a). This effect appears specific for hypercholesterolemia because L-arginine has no such effect in either the microcirculation of normal animals or healthy humans. Interestingly, this effect seems most pronounced in the microcirculation. In one study from our own laboratory, we were unable to demonstrate any effect of L-arginine on abnormal endothelium-dependent relaxations in rabbit aorta (36b). In another study, L-arginine only partially corrected endothelium-dependent relaxations in the basilar artery of hypercholesterolemic rabbits. This discrepancy between large and small vessels was also demonstrated in a recent clinical study by Drexler and co-workers who showed that L-arginine improved flow responses to intracoronary injections of acetylcholine in hypercholesterolemic patients, while having no effect on the abnormal responses of the epicardial coronary arteries (11a). The mechanisms by which L-arginine might improve endothelium-dependent vasodilation in hypercholesterolemia remains unclear. The simplest explanation is that there exists a deficiency of arginine within the endothelium of hypercholesterolemic individuals. Reduced plasma levels of L-arginine have recently been documented in hypercholesterolemic patients (28a). The degree of arginine deficiency necessary to impair endothelium-dependent vasodilation would have to be severe, because the intracellular concentrations of arginine are, in general, 1000 times that necessary for activity of NO synthase (45a). Nevertheless, it is now clear that there are unique biochemical pathways within the endothelium responsible for the conversion of citrulline to arginine, and it is possible that abnormalities in this pathway may be induced by hypercholesterolemia (24a).

Enhanced oxidative degradation of the ED(NO)

Even before it was recognized that the endothelium-derived relaxing factor was a nitric oxide-like compound, it was known that the endothelium-derived relaxing factor could be inactivated by the superoxide anion (19, 45). The endothelium generates superoxide anions under basal conditions and may be stimulated to generate excess superoxide anions by several interventions. Implication of excess destruction of the ED(NO) by superoxide anion in the setting of hypercholesterolemia is based on data from our own laboratory. We have found that the endothelium of cholesterol fed rabbits produces excess quantities of nitrogen oxides with minimal vasodilator activity (36a). This finding is compatible with the concept that the ED(NO) is inactivated before being released from the endothelial cell. In a subsequent study, we found that 1-week treatment with polyethylene-glycolated superoxide dismutase markedly enhanced endothelium-dependent vascular relaxations to acetylcholine and the calcium ionophore A23187 in cholesterol fed, but not normal rabbits (Fig. 6) (36c). The most likely explanation for this finding is that in the atherosclerotic vessel, excess generation of oxygen derived radicals lead to destruction of the EDRF before it reaches the vascular smooth muscle and this could be prevented by increasing the vascular concentration of superoxide dismutase. In preliminary experiments, we have found that the production of superoxide anion is increased in the aortic endothelium of hypercholesterolemic rabbits. The source of excess superoxide anion production by the endothelium remains unknown.

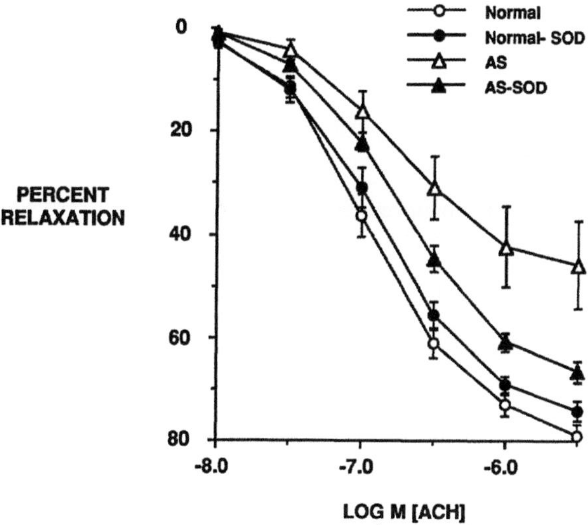

Fig. 6. Effect of treatment with poly-ethylene glycolated superoxide dismutase (PEG-SOD) on endothelium-dependent relaxations to acetylcholine in normal and cholesterol-fed rabbits. Rabbits were fed either a normal diet or a 1 % cholesterol diet for 4 months. During the week prior to study, one-half of each group of animals was treated with PEG-SOD (41,000 U/kg/day I.M.). The vessels were studied in isolated organ chambers as ring segments. (Reproduced with permission from the American Physiological Society and the authors.)

97

Summary and implications

While it has been possible in both experimental animals and humans to demonstrate impaired endothelium-dependent modulation of vascular smooth muscle in the setting of atherosclerosis, the significance of this phenomenon in terms of human disease has yet to be clearly defined but is likely important. Recently, it has been shown that the production of endothelium-derived constricting factors increases with age (31). Since ED(NO) may play an important role in the maintenance of normal blood pressure (41), it is conceivable that the loss of ED(NO) with atherosclerosis, together with the increased production of constrictor factors with age, may predispose to "essential" hypertension, which so commonly develops in the elderly.

Loss of ED(NO) effect may unmask or enhance constrictor effects of both humoral and locally produced substances (some released by subintimal inflammatory cells). This may contribute to several syndromes of altered vasomotion including unstable angina, myocardial infarction, variant angina encountered in the setting of atherosclerosis. The ED(NO) also inhibits platelet aggregation (1, 4). It is interesting to speculate that the loss of this effect may enhance the propensity for local platelet deposition leading to vascular occlusion and the release of both vasoconstrictor factors and factors that may accelerate the development of atherosclerosis. Finally, it is now apparent that the effect of atherosclerosis on the endothelium is not limited to the large vessels but also occurs in the coronary microcirculation. Thus, in addition to predisposing to altered vasomotion of large vessels, endothelial dysfunction in atherosclerosis may dramatically impair neurohumoral regulation of tissue perfusion at the microvascular level.

References

1. Alheid U, Reichwehr I, Förstermann U (1989) Human endothelial cells inhibit platelet aggregation by separately stimulating platelet cyclic AMP and cyclic GMP. Eur J Pharmacol 164: 103–110
2. Bassenge E, Busse R, Pohl U (1987) Abluminal release and asymmetrical response to the rabbit arterial wall to endothelium-derived relaxing factor. Circ Res 61: II-68–II-73
3. Bossaller C, Habib GB, Yamamoto H, Williams C, Wells S, Henry PD (1987) Impaired muscarinic endothelium-dependent relaxation and cyclic guanosine 5'-monophosphate formation in atherosclerotic human coronary artery and rabbit aorta. J Clin Invest 79: 170–174
4. Busse R, Luckhoff A, Bassenge E (1987) Endothelium-derived relaxant factor inhibits platelets activation. Arch Pharmacol 336: 566–577
5. Cherry PD, Furchgott RF, Zawadski JV, Jothianandan D (1982) Role of endothelial cells in relaxation of isolated arteries by bradykinin. Proc Natl Acad Sci USA 79: 2106–2110
5a. Chilian WM, Eastham CL, Marcus ML (1986) Microvascular distribution of coronary vascular resistance in beating left ventricle. Am J Physiol 251 (4 pt 2): H779–788
6. Cocks TM, Angus AR (1983) Endothelium-dependent relaxation of coronary arteries by noradrenaline and serotonin. Nature 305: 627–630
7. Cohen RA, Shepherd JT, Vanhoutte PM (1983) Inhibitory role of the endothelium in response of isolated coronary arteries to platelets. Science 221: 273–274

7a. Cooke JP, Andon NA, Girerd XJ, Hirsch AT, Creager MA (1991) Arginine restores cholinergic relaxation of hypercholesterolemic rabbit thoracic aorta. Circulation 83 (3): 1057–1062

8. Cox DA, Vita JA, Treasure CB, Fish RB, Alexander RW, Ganz P, Selwyn AP (1989) Atherosclerosis impairs flow-mediated dilation of coronary arteries in man. Circulation 80: 458–465

9. De Mey JG, Claeys M, Vanhoutte PM (1982) Endothelium-dependent inhibitory effects of acetylcholine, adenosine triphosphate, thrombin and arachidonic acid in the canine femoral artery. J Pharmacol Exp Ther 222: 166–173

10. De Mey JG, Vanhoutte PM (1981) Role of the intima in cholinergic and purinergic relaxation of isolated canine femoral arteries. J Physiol (Lond.) 316: 347–355

11. Drexler H, Zeiher AM, Wollschlager H, Meinertz T, Just H, Bonzel T (1989) Demonstration of flow-dependent coronary dilatation in man. Circulation 80: 466–474

11a. Drexler H, Zeiher AM, Meinzer K, Just H (1991) Correction of endothelial dysfunction in coronary microcirculation of hypercholesterolaemic patients by L-arginine. Lancet 338 (8782–8783): 1546–1550

12. Dzau VJ (1986) Significance of the vascular renin-angiotensin pathway. Hypertension 8: 553–559

13. Feletou M, Vanhoutte PM (1988) Endothelium-dependent hyperpolarization of canine coronary smooth muscle. Br J Pharmacol 93: 515–524

13a. Flavahan (1992) Atherosclerosis or Lipoprotein-Induced Endothelial Dysfunction. Potential Mechanisms Underlying Reduction in EDRF/Nitric Oxide Activity. Circulation 85 (5): 1927–1938

14. Förstermann U, Mügge A, Alheid U, Haverich A, Frölich JC (1988) Selective attenuation of endothelium-mediated vasodilation in atherosclerotic human coronary arteries. Circ Res 62: 185–190

15. Förstermann U, Trogisch G, Busse R (1984) Species-dependent differences in the nature of endothelium-derived vascular relaxing factor. Eur J Pharmacol 106: 639–643

16. Freiman PC, Mitchell GC, Heistad DD, Armstrong ML, Harrison DG (1986) Atherosclerosis impairs endothelium-dependent vascular relaxation to acetylcholine and thrombin in primates. Circ Res 58: 783–389

17. Furchgott RF, Zawadski JV (1980) The obligatory role of endothelial cells in the relaxation of arterial smooth muscle by acetylcholine. Nature 228: 373–376

17a. Girerd XJ, Hirsch AT, Cooke JP, Dzau VJ, Creager MA (1990) L-arginine augments endothelium-dependent vasodilation in cholesterol-fed rabbits. Circ Res 67 (6): 1301–1308

18. Griffith TM, Edwards DH, Lewis MJ, Newby AC, Henderson HA (1984) The nature of endothelium derived vascular relaxant factor. Nature 308: 645–647

19. Gryglewski RJ, Moncada S, Palmer RMJ (1986) Bioassay of prostacyclin and endothelium-derived relaxing factor (EDRF) from porcine aortic endothelial cells. Br J Pharmacol 87: 685–694

20. Gryglewski RJ, Palmer RMJ, Moncada S (1986) Superoxide anion is involved in the breakdown of endothelium-derived vascular relaxing factor. Nature 320: 454–456

21. Guerra R Jr, Brotherton AFA, Goodwin PJ, Clark CR, Armstrong ML, Harrison DG (1989) Mechanisms of abnormal endothelium-dependent vascular relaxation in atherosclerosis: Implications for altered autocrine and paracrine function of EDRF. Blood Vessels 26: 300–314

22. Harrison DG, Armstrong ML, Freiman PC, Heistad DD (1987) Restoration of endothelium-dependent relaxation by dietary treatment of atherosclerosis. J Clin Invest 80: 1808–1811

23. Harrison DG, Freiman PC, Armstrong ML, Marcus ML, Heistad DD (1987) Alterations of vascular reactivity in atherosclerosis. Circ Res 61: II-74–II-80

24. Habib JB, Bossaller C, Wells S, Williams C, Morrisett JD, Henry PD (1986) Preservation of endothelium-dependent vascular relaxation in cholesterol-fed rabbit by treatment with the calcium blocker PN 200110. Circ Res 58: 305–309

24a. Hecker M, Sessa WC, Harris HJ, Anggard EE, Vane JR (1990) The metabolism of L-arginine and its significance for the biosynthesis of endothelium-derived relaxing factor: cultured endothelial cells recycle L-citrulline to L-arginine. Proc natl Acad Sci USA 87 (21): 8612–8616

25. Holtz J, Förstermann U, Pohl U, Giesler M, Bassenge E (1984) Flow-dependent, endothelium-mediated dilation of epicardial coronary arteries in conscious dogs. Effects of cyclo-oxygenase inhibition. J Cardiovasc Pharmacol 6: 1161–1169

26. Holtz J, Giesler M, Bassenge E (1983) Two dilatory mechanisms of anti-anginal drugs on epicardial coronary arteries in-vivo: Indirect, flow-dependent, endothelium-mediated dilation and direct smooth muscle relaxation. Z Cardiol 72: 98–106

27. Inou T, Tomoike H, Hisano K, Nakamura M (1988) Endothelium determines flow-dependent dilation of the epicardial coronary in dogs. J Am Coll Cardiol 11: 187–191

28. Jayakody L, Senaratne M, Thomson A, Kappagoda T (1987) Endothelium-dependent relaxation in experimental atherosclerosis in the rabbit. Circ Res 60: 251–264

28a. Jeserich M, Münzel T, Just H, Drexler H (1992) Reduced plasma L-arginine in hypercholesterolaemia. Lancet 339: 561

29. Katusic ZS, Shepherd JT, Vanhoutte PM (1984) Vasopressin causes endothelium-dependent relaxations of the canine basilar artery. Circ Res 55: 576–579

30. Katusic ZS, Vanhoutte PM (1989) Superoxide anion is an endothelium-derived contracting factor. Am J Physiol 257: H33–H37

31. Koga T, Takata Y, Kobayashi K, Takishita S, Yamashita Y, Fujishima M (1989) Age and hypertension promote endothelium-dependent contractions to acetylcholine in the aorta of the rat. Hypertension 14: 542–548

31a. Kuo L, Davis MJ, Cannon MS, Chilian WM (1992) Pathophysiological consequences of atherosclerosis extend into the coronary microcirculation. Restoration of endothelium-dependent responses by L-arginine. Circ Res 70 (3): 465–476

32. Lopez JA, Armstrong ML, Harrison DG, Piegors DJ, Heistad DD (1989) Vascular responses to leukocyte products in atherosclerotic primates. Circ Res 65: 1078–1086

33. Ludmer PL, Selwyn AP, Shook TL, Wayne RR, Mudge GH, Alexander RW, Ganz P (1986) Paradoxial vasoconstriction induced by acetylcholine in atherosclerotic coronary arteries. N Engl J Med 315: 1046–1051

34. McLenachan JM, Williams JK, Ganz P, Selwyn AP (1989) Loss of flow-mediated endothelium-dependent dilation in atherosclerosis. Circulation 80: II-279

35. Miller VM, Vanhoutte PM (1989) Is nitric oxide the only endothelium-derived relaxing factor in canine femoral veins? Am J Physiol 26: H1910–1916

36. Miller VM, Vanhoutte PM (1985) Endothelium-dependent contractions to arachidonic acid are mediated by products of cyclooxygenase in canine veins. Am J Physiol 248: H432–H437

36a. Minor RLJ, Myers PR, Guerra RJ, Bates JN, Harrison DG (1990) Diet-induced atherosclerosis increases the release of nitrogen oxides from rabbit aorta. J Clin Invest 86: 2109–2116

36b. Mügge A, Harrison DG (1991) L-arginine does not restore endothelial dysfunction in atherosclerotic rabbit aorta in vitro. Blood Vessels 28: 354–357

36c. Mügge A, Elwell JH, Peterson TE, Hofmeyer TG, Heistad DD, Harrison DG (1991) Chronic treatment with polyethylene glycolated superoxide dismutase partially restores endothelium-dependent vascular relaxations in cholesterol-fed rabbits. Circ Res 69: 1293–1300

37. Myers PR, Minor RL Jr., Guerra R Jr., Bates JN, Harrison DG (1990) Vasorelaxant properties of the endothelium-derived relaxing factor more closely resemble S-nitrosocysteine than nitric oxide. Nature (Lond.) 345: 161–163

38. Okumura K, Yasue H, Matsuyama K, Goto K, Miyagi H, Ogawa H (1988) Sensitivity and specificity of intracoronary injection of acetylcholine for the induction of coronary artery spasm. J Am Coll Cardiol 12: 883–888

39. Palmer RMJ, Ferrige AG, Moncada S (1987) Nitric oxide release accounts for the biological activity of endothelium-derived relaxing factor. Nature (Lond.) 327: 524–526

40. Piper GM, Gross GJ (1989) Selective impairment of endothelium-dependent relaxation by oxygen-derived free radicals: Distinction between receptor versus nonreceptor mediators. Blood Vessels 26: 44–47

40a. Quillen JE, Sellke FW, Armstrong ML, Harrison DG (1991) Long-term cholesterol feeding alters the reactivity of primate coronary microvessels to platelet products. Arteriosclerosis and Trombosis 11: 639–644

41. Rees DD, Palmer RMJ, Moncada S (1989) Role of endothelium-derived nitric oxide in the regulation of blood pressure. Proc Natl Acad Sci USA 86: 3375–3378

42. Reiko T, Werb Z (1984) Secretory products of macrophages and their physiological function. Am J Physiol 246: C1–C9

42a. Rossitch E Jr, Alexander E, Black PM, Cooke JP (1991) L-arginine normalizes endothelial function in cerebral vessels from hypercholesterolemic rabbits. J Clin Invest 87: 1295–1299

43. Rubanyi GM, Romero JC, Vanhoutte PM (1986) Flow-induced release of endothelium-derived relaxing factor. Am J Physiol 250: H145–H149

44. Rubanyi GM, Vanhoutte PM (1987) Nature of endothelium-derived relaxing factor: Are there two relaxing mediators? Circ Res 61: II-61–II-67

45. Rubanyi GM, Vanhoutte PM (1986) Oxygen-derived free radicals, endothelium and responsiveness of vascular smooth muscle. Am J Physiol 250: H815–H821

45a. Schmidt HHHW, Pollock JS, Nakane M, Gorsky L, Förstermann U, Murad F (1991) Purification of a soluble isoform of guanylyl cyclase-activating-factor synthase. Proc Natl Acad Sci USA 88: 365–369

46. Sellke FW, Armstrong ML, Harrison DG (1990) Endothelium-dependent vascular relaxation is abnormal in the coronary microcirculation of atherosclerotic primates. Circulation 81: 1586–1593

47. Shimokowa H, Aarhus LL, Vanhoutte PM (1987) Porcine coronary arteries with regenerated endothelium have a reduced endothelium-dependent responsiveness to aggregating platelets and serotonin. Circ Res 61: 256–270

48. Shimokawa H, Vanhoutte PM (1988) Dietary cod-liver oil improves endothelium-dependent responses in hypercholesterolemic and atherosclerotic porcine coronary arteries. Circulation 78: 1421–1430

49. Shirahase H, Usui H, Kurahashi K, Fujiwara M, Fukui K (1988) Endothelium-dependent contraction induced by nicotine in isolated canine basilar artery: possible involvement of a thromboxane A_2 (TXA_2) like substance. Life Sci 42: 437–445

50. Tesfamarian B, Jakubowski JA, Cohen RA (1989) Contraction of diabetic rabbit aorta caused by endothelium-derived PGH_2-TxA_2. Am J Physiol 257: H1327–H1333

51. Toda N (1986) Mechanisms of histamine-induced relaxation in isolated monkey and dog coronary arteries. J Pharmacol Exp Ther 239: 529–535

52. Vanhoutte PM, Houston DS (1985) Platelets, endothelium and vasospasm. Circulation 72: 728–734

53. Vekshtein VI, Yeung AC, Vita JA, Nabel EG, Fish RD, Bittl JA et al. (1989) Fish oil improves endothelium-dependent relaxation in patients with coronary artery disease. Circulation 80: II-434

54. Verbeuren TJ, Jordaens FH, Zonnekeyn LL, Van Hove CE, Coene MC, Herman AG (1986) Effect of hypercholesterolemia on vascular reactivity in the rabbit. Circ Res 58: 552–564

55. Vita JA, Treasure CB, Nabel EG, Fish RD, McLenachan JM, Yeung AC, Vekshtein VI et al. (1989) The coronary response to acetylcholine relates to coronary risk factors. Circulation 80: II-435

55a. Vita JA, Treasure CB, Nabel EG, McLenachan JM, Fish RD, Yeung AC, Vekshtein VI, Selwyn AP, Ganz P (1990) Coronary vasomotor response to acetylcholine relates to risk factors for coronary artery disease. Circulation 81 (2): 491–497

56. Warner TD, Mitchell JA, de Nucci G, Vane JR (1989) Endothelin-1 and endothelin-3 release EDRF from isolated perfused arterial vessels of the rat and rabbit. J Cardiovasc Pharmacol 13: S85–S88
57. Yamamoto H, Bossaller C, Cartwright J Jr., Henry PD (1988) Videomicroscopic demonstration of defective cholinergic arteriolar vasodilation in atherosclerotic rabbit. J Clin Invest 81: 1752–1758
58. Yanagisawa M, Kurihara H, Klimura S, Tomobe Y, Kobayashi M, Mitsui Y, Yazaki Y, Goto K, Masaki T (1988) A novel potent vasoconstrictor peptide produced by vascular endothelial cells. Nature 332: 411–415

Author's address:
David G. Harrison
Cardiology Division
P.O. Box LL
Emory University School of Medicine
Atlanta, GA 30322
USA

New determinants of the uptake of atherogenic plasma proteins by arteries

G. V. R. Born

The William Harvey Research Institute, St. Bartholomew's Hospital Medical College, Charterhouse Square, London, England

Summary: Atherosclerotic disease begins with the accumulation of atherogenic plasma proteins, predominantly low-density lipoprotein (LDL), in susceptible arteries. One determinant of this is the LDL concentration in the blood (4). We are investigating other factors which may influence the uptake of LDL and of fibrinogen, another atherogenic plasma protein, by artery walls, as well as the mechanism(s) of this uptake. In order to model human atherogenesis as closely as possible, the experiments are mostly in vivo, using rabbits and rats.

In anesthetized rabbits, intravenous infusions of histamine sufficient to produce inter-endothelial gaps throughout the venular systems did not increase the rate constant of the disappearance of LDL from the circulating blood. This makes it improbable that significant quantities of LDL pass out of the blood into vessel walls between endothelial cells, and it indirectly supports the evidence that LDL leaves the plasma by transcytosis through the endothelial cells.

In anesthetized rabbits and in conscious rats the uptake of LDL, methylated to prevent its removal by high-affinity receptors (m-LDL), is significantly increased by noradrenaline and by adrenaline at their pathophysiological blood concentrations. In man this could help to account for the coronary risk factor status of cigarette smoking, hypertension, stress, etc. when associated with increases in circulating catecholamines. The mechanism of this effect is now under investigation.

Key words: Atherogenesis − low density lipoprotein − catecholamines − endothelium

Evidence against the atherogenic passage of LDL between endothelial cells

There is convincing evidence that LDL passes from the blood plasma into arterial walls predominantly through the cells of the endothelial layer (9, 10, 11). This "transcytosis" appears to involve the plasmalemmal vesicles which are characteristic of endothelium; but the actual mechanism is far from being understood. The question remained whether significant amounts of LDL are able also to pass through the junctions between endothelial cells. If so, it might be expected that such a passage would become demonstrable in vivo as increases in the blood clearance rates of LDL, methylated to prevent clearance by the high affinity receptors (particularly in the liver), under conditions in which the junctions become separated to form wide inter-endothelial gaps throughout a major part of the circulation, i.e. throughout the venular systems. This can be produced by the inflammatory mediator histamine (6).

We have tested the idea as follows (1): Unanaesthetized restrained rabbits were injected with human LDL (which for these experiments could be considered sufficiently similar to rabbit LDL), which had been methylated and labelled with radioiodine. By measuring the radioactivity of blood samples taken at frequent intervals over periods of several hours and calculating lines of best fit, rate constants of the apparent clearance of m-LDL can be determined with remarkable accuracy, so that even small changes in such rates become evident. Blood samples were taken from a catheter in the central artery of one ear for determinations of radioactivity and haematocrit every 20 min during successive three hour periods of infusion (at a rate of 0.05 ml/min) into the marginal vein of the other ear; first of saline, then of histamine (12.5 ug/kg.min) and finally again of saline. The histamine concentrations were chosen to have no significant effect on the heart rate or blood pressure while producing in the venules wide inter-endothelial gaps which were demonstrable morphologically. During histamine infusions the apparent LDL clearance rates, instead of increasing, actually decreased. At the same time the histamine infusions produced the well-established increases in the haematocrit, due to loss of plasma water through the gaps, with associated haemoconcentration. This haemoconcentration accounted accurately for the apparent decrease in the LDL clearance rate, so that the m-LDL concentration in the plasma increased to exactly the same extent as the red cell concentration. Unless the structure of the intimal tissues underlying the endothelium is very different in veins and arteries our results make it improbable that significant quantities of LDL pass out of the blood into vessel walls between endothelial cells, and thus give indirect support to the transcytotic pathway in atherogenesis.

Acceleration by catecholamines of the atherogenic uptake of low density lipoproteins by arteries.

There is very little knowledge as to whether and, if so, how endogenous agents influence atherogenesis. Specifically, it appeared to be unknown whether the catecholamines influence the removal of LDL from the blood or the accumulation of LDL in arterial walls. We have begun to investigate this question, with promising results (2, 8).

We discovered that in anaesthetized rabbits the atherogenic uptake of LDL by arterial walls is accelerated by noradrenaline and by adrenaline at their pathophysiological concentrations in rabbit and human blood. The principle of the experiments was to compare the uptake of intravenously injected, radioactively labelled LDL, methylated to prevent removal by high-affinity receptors, in the two carotid arteries of anaesthetized rabbits after infusing low concentrations of noradrenaline or of adrenaline into one carotid and saline as control into the other, the volume rates of infusion being about one per cent of the carotid blood flows. Human LDL, which behaves sufficiently like rabbit LDL for these purposes, was prepared, methylated and radio-iodinated by standard methods. At the end of the infusions, the arteries were excised and their radioactivities determined. Noradrenaline infused for two h to produce local blood concentrations of nominally, 1, 10, 50 and 100 nM significantly increased the LDL radioactivities of the walls of the noradrenaline-infused carotids. Concentrations of nominally 100 nM also increased the LDL radioactivities of the walls of the saline-infused carotids; this was associated with significant increases in their blood noradrenaline concentrations (kindly determined by Prof. M. Brown, Department of Clinical Pharmacology in Cambridge).

We have recently demonstrated with a different technique (3) that this effect also occurs with adrenaline in another species, i.e. the rat.

Osmotic minipumps (ALZET) were implanted in rats under the skin of the neck and connected to the left common carotid artery. These pumps infused either saline or adrenaline at ca. 1 μl/h to give plasma adrenaline concentrations of ca. 40 or 200 nM for 6 days in unrestrained conscious rats.

Human and rat LDL prepared by sequential ultracentrifugation (5) were labelled with ^{125}I-tyramine cellobiose (^{125}I-TC-LDL); when the lipoprotein itself undergoes degradation and removal, the labelled adduct remains in the tissue to indicate how much LDL has been there (7). After 5 days' infusion, ^{125}I-TC-LDL was injected i.v., with the infusion continued, and 24 h later the animals were killed with intra-cardiac pentobarbital and the thoracic aortae prepared (8) for radioactivity determinations.

Adrenaline infused at either concentration increased the aortic radioactivities due to human LDL by 146 % (n = 11) and due to rat LDL by 289 % (n = 7); these inceases were both highly significant.

The results confirmed that the catecholamines at high pathophysiological blood concentrations accelerate the uptake of LDL by large arteries in two mammalian species, viz. rabbit and rat.

These results strongly suggest that the effect is also present in man. This discovery, therefore, appears to contribute towards an explanation for the accelerated atherosclerosis and the increased incidence of its clinical manifestations in conditions associated with elevated blood catecholamine concentrations, including the episodic increases associated with stress and cigarette smoking as well as the more persistent increases caused by phaeochromocytoma. Thus, there is the possibility that the slow but continuous passage of LDL into arterial walls depends, at least in part, on the catecholamines normally present in the circulating blood.

The mechanism underlying the accelerating effect of the catecholamines on m-LDL uptake remains to be established. The possibility has to be considered that it is due to their hypertensive effect, which could also account for the accelerated atherogenesis associated with pheochromocytoma. The local blood pressure was not significantly increased by noradrenaline infused at nominally 1 and 10 nM and only slightly incresed by 100 nM. These experiments, therefore, do not provide evidence on this question.

The catecholamine effect clearly does not involve the high-affinity receptors which, although present on vascular endothelium (12) do not bind or remove the methylated LDL used here. Another possible route for the movement of LDL from blood into arterial walls is through inter-endothelial junctions. However, this route has been made very unlikely by our experiments, described above, in which the clearance of LDL from the blood increased not at all when wide inter-endothelial gaps were induced pharmacologically in a large proportion of the rabbit vasculature (1). This leaves as the predominant if not the only route the passage of LDL by transcytosis through endothelial cells in plasmalemmal vesicles (9, 10). It is therefore reasonable to suggest that it is this process that is somehow accelerated by the catecholamines. The possibility of antagonizing it pharmacologically may provide a new way of slowing the progress of atherosclerosis.

The accelerated LDL uptake here described is brought about by circulating catecholamines in direct and continuous contact with the arterial endotelium where it presumably mediates the effect. Therefore our findings raise the possibility that, as well as the catecholamines, other circulating mediators affect the flux of LDL and of other plasma proteins into arterial walls; these possibilities are being looked into.

Acknowledgements:

This work was done with Drs. Peter Görög, Shahida Shafi and Eduardo Cardona-Sanclemente.

Generous support from the British Heart Foundation and the Garfield Weston Foundation is gratefully acknowledged.

References

1. Born GVR, Shafi S (1989) Evidence against the atherogenic passage of low-density lipoproteins between vascular endothelial cells in conscious rabbits. J Physiol Lond, 418, 80P
2. Born GVR, Shafi S, Cusack NJ (1989) Evidence for the acceleration of atherogenesis by circulating epinephrine. Transplantation Prof, 21, 3660–3661
3. Cardona-Sanclemente EL, Görög P, Born GVR (1992) Increased uptake of low-density lipoprotein induced by adrenaline in the aortae of unrestrained rats. J Physiol Lond (in press)
4. Goldstein JL, Brown MS (1977) The low-density lipoprotein pathway and its relation to atherosclerosis. Annu Rev Biochem, 897–930
5. Havel RG, Edner HA, Bragdon G (1955) The distribution and chemical composition of ultra centrifugally separated lipoproteins in human serum. J Clin Invest 34, 1345
6. Majno G, Shea SM, Leventhal M (1969) Endothelial contraction induced by histamine – type mediators: an electron-microscope study. J Cell Biol 42: 647–672
7. Pittman RC, Carew TE, Glass CK, Green SR, Taylor CA, Attie AD (1983) Radioudinated, intracellularly trapped ligand for determining the sites of plasma protein degradation in vivo. Biochem Journ 212, 791
8. Shafi S, Cusack NJ, Born GVR (1989) Increased uptake of methylated low density lipoprotein induced by noradrenaline in carotid arteries of anaesthetised rabbits. Proc Roy Soc Lond B 235: 289–293
9. Simionescu N, Vasile E, Lupu F, Popescu G, Simionescu M (1986) Prelesional events in atherogenesis accumulation of extracellular cholesterol-rich liposomes in the arterial intima and cardiac valves of the hyperlipidaemic rabbit. Am J Pathol 123: 109–125
10. Stein O, Stein Y, Eisenberg S (1973) A radiographic study of the transport of I-labelled serum lipoproteins in rat aorta. Cell Tissue Res 138: 223–232
11. Vasile E, Simionescu M, Simionescu N (1983) Visualization of the binding, endocytosis, and transcytosis of low-density lipoprotein in the arterial endothelium in situ. J Cell Biol 96: 1677–1689
12. Vlodavsky I, Fielding PE, Fielding CJ, Gospadorowicz D (1978) Role of contact inhibition in the regulation of receptor-mediated uptake of low density lipoprotein in cultured vascular endothelial cells. Proc Natl Acad Sci, USA, 75: 356–360

Author's address:
G. V. R. Born
The William Harvey Research Institute
St. Bartholomew's Hospital Medical College
Charterhouse Square
London, EC1M 6BQ
England

106

Hyperlipidemic endothelial injury and angiogenesis

P. D. Henry

Baylor College of Medicine, Department of Medicine, Houston, Texas, USA

Summary: Hypercholesterolemia is associated with endothelial cell dysfunction which may in part be related to an accumulation of toxic lipoprotein degradation products in artery walls. Oxidized low-density lipoprotein (LDL) and its products have been incriminated in impairing various endothelial functions including G-protein-dependent transmembrane signaling, calcium regulation, phosphoinositide turnover, protein kinase C activation and others. Modification of such cell regulatory functions may alter the responsiveness of endothelial cells to angiogenic (mitogenic) stimuli. Endothelial cell replication is necessary for the growth of pre-existing arterial channels and the formation of new microvessels (angiogenesis). Experiments in intact rabbits indicate that endothelial replication necessary for vascular growth is markedly impaired in the presence of hypercholesterolemia, a defect that could play an important role in the pathophysiology of occlusive atherosclerotic disease.

Key words: Oxidized LDL – lysophosphatidylcholine – endothelial cell – cytokines – angiogenesis

Introduction

There is general agreement that atherogenic dyslipidemic states are associated with changes in the structure and function of endothelial cells (33). Studies on altered endothelial function in atherosclerosis have addressed predominantly changes affecting endothelial vasomotor regulation, endothelial barrier function, endothelial interactions with blood cells (leucocytes, platelets) and regulation of thrombosis and fibrinolysis. On the other hand, one of the most important functions of endothelial cells, their participation in vessel growth (angiogenesis), has received suprisingly little attention. Because endothelial cells in the presence of dyslipidemia appear to exhibit multiple functional impairments, the question of their capacity to undergo appropriate angiogenic growth responses is of interest. The pathophysiological impact of occlusive arterial disease strongly depends upon compensatory arterial growth (collateral formation).

Early work on endothelial changes in atheroslcerosis focused primarily on gross structural changes (endothelial desquamation) (33, 50), but more recently, attempts have been made to characterize endothelial changes in molecular terms. Because atherosclerotic responses are evoked by high fat (cholesterol) diets, it appears logical to look at alterations in lipid metabolism as a pathogenic mechanism of endothelial functional injury.

Hyperlipidemic endothelial injury

In 1952, Glavind et al. (12) demonstrated for the first time an accumulation of peroxidized lipids in the walls of atherosclerotic arteries and "suggested the possibility of an active part played by these compounds in the pathogenesis of atherosclerosis". Subsequently, Harman (14) in 1957 summarized atherosclerosis as resulting from "three primary processes: 1) oxidative polymerization of constituents of serum lipoproteins; 2) anchoring of the oxidized material in the arterial wall; 3) an inflammatory reaction induced in the arterial wall by these oxidation products". In the late 1970s, Henriksen et al. (15, 16) and Hessler et al. (19) demonstrated that preparations of isolated human low-density lipoprotein (LDL; abbreviations in this paper are listed in legend to Table 1) exerted toxic effects on endothelial cells in culture. In addition, Henriksen in collaboration with Mahoney and Steinberg (17) showed that LDL exposed to endothelial cells in culture underwent a chemical modification that greatly enhanced its susceptibility to uptake and degradation by ma-

Table 1. Effects of oxidatively modified LDL and lysophosphatidylcholine on endothelial function

	Oxidized LDL	LPC	References
Impairment of endothelium-dependent arterial relaxation	+	+	3, 27, 32, 47
Modulation of G-protein function	+	+	23
Stimulation of release of internal Ca (transient effect)	+	+	20
Inhibition of agonist-induced release of Ca and Ca-influx (sustained effect)	0	+	20, 36
Activation of phospho-inositide turnover	+	+	20, 21
Activation of protein kinase C activity	+	+	26
Activation of tissue factor activity and reduction of protein C activity	+	0	8, 51
Stimulation of expression of plasminogen activator inhibitor-1 (PAI-1)	+	0	29
Activation of release of cytokines MCP-1 and M-CSF	+	0	7, 44
Inhibition of the release of PDGF	+	0	9
Stimulation of expression of leucocyte adhesion molecules (VCAM-1, and ICAM-1)	0	+	28
Impairment of endothelial replication and cytotoxic effects	+	+	15, 16, 19, 24, 36

0: No data available; **LDL:** low density lipoprotein; **LPC:** lysophosphatidylcholine; **PAI-1:** plasminogen activator inhibitor 1; **MCP-1:** macrophage chemotactic protein 1; **VCAM-1:** vascular cell adhesion molecule 1; **ICAM-1:** intercellular adhesion molecule 1.

crophages. Morel et al. (34) and Steinbrecher et al. (48) were then able to show that the chemical modification of native LDL after incubation with cells in culture was related to a peroxidation of unsaturated fatty acids in LDL. Steinbrecher et al. (48) also demonstrated that peroxidative modification of LDL was associated with a substantial degradation of phosphatidylcholine to authentic lysophosphatidylcholine (LPC). Further studies suggested that the generation of LPC in oxidatively modified LDL was related to a phospholipase A_2 activity closely associated with LDL (39, 40, 42). Lipid peroxidation products reacting with lysine residues of apo B in LDL confers new immunogenic properties to the apoprotein (38). In support of the concept that lipoprotein oxidation products accumulate in vivo in artery walls exposed to hyperlipidemia is the finding that atheromatous lesions bind monoclonal antibodies specific for oxidatively modified apo B (45).

In general, oxidized LDL and one of its important lipid constituents, lysophosphatidylcholine (lysolecithin, LPC), exert anti-vasodilator (pro-constrictor), pro-inflammatory, and pro-thrombotic (antifibrinolytic) effects on endothelial cells (Table 1). Oxidized LDL may exert modulatory effects on other cells involved in atherogenesis. For instance, effects on signaling pathways, cell activation, and cytokine release have been reported for vascular smooth muscle cells (43), monocytes (10, 22, 25, 28, 31), and T-lymphocytes (11). Although lysolipids may account for some of the actions of oxidized LDL, it is clear that other molecular effectors including modified apo B, hydroperoxides, aldehydes, and oxysterols may also be important (2, 25). LPC or oxidized LDL has often shown concentration-dependent effects in which low agonist concentration exert stimulatory and high concentrations inhibitory effects (23, 37).

Repercussion of endothelial dysfunction on mechanisms of angiogenesis

As shown in Table 1, oxidized LDL and LPC alter pathways which are known to play a pivotal role in cell replication (20, 21, 26, 37, 49). Important pathways include those affecting G-protein-dependent transmembrane signaling (23, 47), phospholipid (phosphoinositide) turnover (20, 21, 26), calcium regulation (20, 36), protein kinase C activation (phorbol-like effects) (26, 37), and release of cytokines or growth factors (9, 25, 31). Because of the multiple stimulatory or inhibitory effects of oxidized LDL (or LPC), overall effects of these agents in vivo are difficult to analyze.

Distinction between physiological ("non-inflammatory") and pathological ("inflammatory") angiogenesis

The fact that hypercholesterolemic states are associated with an altered endothelium-dependent regulation of microvessels (arterioles) (vessels not known to develop atheromatous lesions) indicates that endothelial dysfunction is not merely a consequence of lesion formation (18, 52). In the context of angiogenesis, it is important to distinguish between physiological growth responses and growth responses occurring focally at the site of an inflammatory response. Physiological growth of large and small vessels in response to gradual alterations in the balance between metabolic organ demand and supply (organ growth, exercise, decreased oxygen delivery due to high altitude) occurs in the absence of inflammatory responses (leucocytic infiltrates). In contrast, at the site of inflammatory responses, angiogenesis is associated with prominent cellular infiltrates. It has been generally recognized that processes involved in the generation of atheromatous lesions are closely related to an inflammatory (or immune) response (13, 30, 35). Thus, during early atherogenesis, tissue sites undergoing atheromatous changes are infiltrated by inflammatory mononuclear cells (monocyte/macrophages, T-lymphocytes) (13, 30). Atheromas have features in common with non-neutrophilic chronic inflammatory responses as seen in foreign body granulomatous reactions (46). The presence of mononuclear infiltrates is important, because these cells are known to release a variety of mitogens and mediators that may markedly amplify angiogenic responses (inflammatory neovascularization) (30). It is not surprising that atheromatous lesions exhibiting many signs of an inflammatory response, including an accumulation of mononuclear inflammatory cells, determine a local angiogenic response leading to a "vasa vasorum" network surrounding the lesions (1). These vasa vasorum networks should not be misinterpreted to mean that vascular growth in dyslipidemic states is somehow systemically increased and would augment vascular growth in response to physiological angiogenic stimuli. Even at the site of atheroma formation. angiogenic responses are difficult to evaluate. It is, for instance, unclear whether vasa vasorum growth is in fact limited by the hyperlipidemic state. The fact that endothelial denudation does occur in atherogenesis does not suggest facilitated endothelial replication at lesion sites (33, 41, 50).

Macrovascular growth model

One way to evaluate effects of hypercholesterolemia on angiogenesis is to generate in intact animals with and without hypercholesterolemia stimuli known to evoke predictable vascular growth responses. Recently, we have performed experiments in cholesterol-fed rabbits subjected to simple vessel growth stimulatory interventions (4, 53).

In one series of experiments, arterial flow through the carotid artery of rabbits with or without hypercholesterolemia was chronically increased by transection of the contralateral carotid artery (53). After this intervention, we were able to demonstrate by high-frequency ultrasound imaging a time-dependent expansion of the lumen of the carotid artery. After 8 weeks, the high flow arteries were shown to have undergone growth responses demonstrable by quantitative morphology and thymidine uptake into intimal (endothelial) and medial (smooth muscle) cells. Interestingly, growth reponses in the hypercholesterolemic group were markedly reduced, although increases in carotid flow immediately after contralateral carotid transection and 8 weeks later were nearly identical in the two groups. Thus, we were unable to demonstrate that reduced carotid growth was related to a limited hyperperfusional response determined by impaired endothelium-dependent relaxation. Of interest was that gentle endothelial denudation of the carotid at the time of contralateral carotid transection produced in normocholesterolemic rabbits striking reductions in carotid growth. Because hypercholesterolemia and mechanical trauma both impair endothelial function, it appears that arterial growth in response to flow stimuli critically depends upon endothelial integrity (53).

Microvascular growth model

In another series of experiments, the distal ear of rabbits with and without hyper-cholesterolemia was subtotally transected (4). This intervention made the perfusion of the distal ear dependent upon a single marginal ear artery (diameter ~275 μm). Appreciable signs of ischemia of the distal ear did not occur after this intervention. In normocholesterolemic rabbits, the auriculotomy produced, after 10 days, a substantial expansion of the marginal artery lumen proximal to the ear cut. No signs of inflammation (no cellular infiltrates of any kind) were demonstrable at any time at the level of the marginal artery proximal to the auriculotomy. The auriculotomy determined in the distal ear no histologic signs of acute ischemia (no neutrophilic infiltrate), although modest mononuclear infiltrates were detected 2–3 days after operation. Microvessels (6–20 μm diameter) in the distal ear of normocholesterolemic rabbits underwent a proliferative response demonstrated by quantitation of microvascular density and radioactive labelling of endothelial nuclei after in vivo pulsing with radioactive thymidine. Marginal ear artery and microvascular proliferations in hypercholesterolemic rabbits were markedly reduced compared to controls. These results confirmed that vascular growth in response to important angiogenic stimuli (hyperperfusion, ischemia/hypoxia) was impaired (4).

In vitro angiogenesis model

To determine whether impaired vascular growth was dependent upon hemodynamic stimuli, organ explant experiments were performed (5, 6). One mm^2 aortic intima-media explants from rabbits with or without hypercholesterolemia were grown in a collagen matrix containing serum-free medium (RPMI 1640) without growth factor supplements. Under these experimental conditions, endothelial cell outgrowth from the edges of the explants was organized as capillary-like microtubes. Videomicroscopic monitoring of the time-dependent elongation of these microtubes was used as an index of endothelial cell growth. Lesion-free explants from hypercholesterolemic rabbits exhibited a strikingly impaired growth. However, supply of exogenous mitogens by co-culturing explants with activated rabbit monocytes nearly corrected impaired growth, which indicated that the endothelial cells were not irreversibly impaired, but remained responsive to mitogens released by activated macrophages (6). These results obtained in an unstirred culture system indicated that impaired growth of hypercholesterolemic endothelium was at least in part independent of fluid mechanical stimuli. Further, they demonstrated the importance of distinguishing growth in the presence or absence of inflammatory cells as encountered in atheromatous lesions.

References

1. Barger AC, Beeuwkes R, Lainey LL, Silverman KJ (1984) Hypothesis: vasa vasorum and neovascularization of human coronary arteries. N Engl J Med 310: 175–177
2. Bhadra S, Arshad MAQ, Rymaszewski Z, Norma E, Wherley R, Subbiah MTR (1991) Oxidation of cholesterol moiety of low density lipoprotein in the presence of human endothelial cells or Cu^{+2} ions: identification of major products and their effects. Biochem Biophys Res Comm 176: 431–440
3. Bossaller C, Habib GB, Yamamoto H, Williams C, Wells S, Henry PD (1987) Impaired muscarinic endothelium-dependent relaxation and cyclic quanosine 5'-monophosphate formation in atherosclerotic human coronary artery and rabbit aorta. J Clin Invest 79: 170–174
4. Bucay M, Nguy JH, Barrios R, Kerns SA, Henry PD (1992) Impaired macro- and microvascular growth in hypercholesterolemic rabbit. J Am Coll Cardiol 19: 151A
5. Chen C-H, Li Z, Kerns SA, Cartwright J, Henry PD (1992) Aortic explants from hypercholesterolemic rabbits exhibit suppressed microvascular growth in an in vitro model of angiogenesis. J Am Coll Cardiol 19: 152A
6. Chen C-H, Li Z, Kerns SA, Cartwright J, Nguy J, Henry PD (1992) Impaired capillary-like endothelial growth of aortic explants from cholesterol-fed rabbits and angiogenic response to foam cells. Circulation 86: I-20
7. Cushing SD, Berliner JA, Valente AJ, Territo MC, Navab M, Parhami F, Gerrity R, Schwartz CJ, Fogelman AM (1990) Minimally modified low density lipoprotein induces monocyte chemotactic protein 1 in human endothelial cells and smooth muscle cells. Proc Natl Acad Sci USA 87: 5134–5138
8. Drake TA, Hannani K, Fei H, Lavi S, Berliner JA (1991) Minimally oxidized low-density lipoprotein induces tissue factor expression in cultured human endothelial cells. Am J Pathol 138: 601–607
9. Fox PL, DiCorleto PE (1986) Modified low density lipoproteins suppress production of a platelet-derived growth factor-like protein by cultured endothelial cells. Proc Natl Acad Sci USA 83: 4774–4778
10. Frostegard J, Nilsson J, Haegerstrand A, Hamsten A, Wigzell H, Gidlund M (1990) Oxidized low density lipoprotein induces differentiation and adhesion of human monocytes and the monocytic cell line U937. Proc Natl Acad Sci USA 87: 904–908
11. Frostegard J, Wu R, Giscombe R, Holm G, Lefvert AK, Nilsson J (1992) Induction of T-cell activation by oxidized low density lipoprotein. Arterisclerosis Thromb 12: 461–467
12. Glavind J, Hartmann S, Clemmesen J, Jessen KE, Dam H (1952) Studies on the role of lipoperoxides in human pathology. II. The presence of peroxidized lipids in the atherosclerotic aorta. Acta Pathol Microbiol Immunol Scand 30: 1–6
13. Hansson GK, Seifert PS, Olsson G, Bondjers G (1991) Immunohistochemical detection of macrophages and T lymphocytes in atherosclerotic lesions of cholesterol-fed rabbits. Arteriosclerosis Thromb 11: 745–750
14. Harman D (1957) Atherosclerosis: a hypothesis concerning the initiating steps in pathogenesis. J Gerontol 12: 199–202
15. Henriksen T, Evensen SA, Carlander B (1977) Injury of endothelial cells in culture induced by low density lipoproteins. Eur J Clin Invest 7: 243
16. Henriksen T, Evensen SA, Carlander B (1979) Injury to cultured endothelial cells induced by low density lipoproteins: protection by high density lipoproteins. Scand J Clin Lab Invest 39: 369–375
17. Henriksen T, Mahoney EM, Steinberg D (1981) Enhanced macrophage degradation of low density lipoprotein previously incubated with cultured endothelial cells: recognition by receptors for acetylated low density lipoproteins. Proc Natl Acad Sci USA 78: 6499–6503
18. Henry PD, Bucay M (1991) Effects of low-density lipoproteins and hypercholesterolemia on endothelium-dependent vasodilation. Curr Opin Lipidol 2: 306–310
19. Hessler JR, Robertson AL, Chisolm GM (1979) LDL-induced cytotoxicity and its inhibition in human vascular smooth muscle and endothelial cells in culture. Atherosclerosis 32: 213–229

20. Hirata K, Akita H, Yokoyama M (1991) Oxidized low density lipoprotein inhibits bradykinin-induced phosphoinositide hydrolysis in cultured bovine aortic endothelial cells. FEBS 287: 181–184

21. Inoue N, Hirata K, Yamada M, Hamamori Y, Matsuda Y, Akita H, Yokoyama M (1992) Lysophosphatidylcholine inhibits bradykin-induced phosphoinositide hydrolysis and calcium transients in cultured bovine aortic endothelial cells. Circ Res 71: 1410–21

22. Jeng JR, Chu CJ, Chang LI, Jan RM, Chiu HC (1991) Oxidized low density lipoprotein enhance monocyte endothelial interaction. FASEB J 5: A1251

23. Kerns S, Mangin E, Nguy J, Henry P (1992) Oxidized LDL and lysolecithin inhibit membrane GTPase-dependent transmembrane signalling. FASEB J 6: A1362

24. Kerns SA, Nguy J, Henry PD (1991) Oxidatively modified low density lipoprotein impairs endothelial growth in vitro. Clin Res 39: 225A

25. Ku G, Thomas CE, Akeson AL, Jackson RL (1992) Induction of interleukin 1 beta expression from human peripheral blood monocyte-derivied macrophages by 9-hydroxy-octadecadienoic acid. J Biol Chem 267: 14183–14188

26. Kugiyama K, Ohgushi M, Sugiyama S, Murohara T, Fukunaga K, Miyamoto E, Yasue H (1992) Lysophosphatidylcholine inhibits surface receptor-mediated intracellular signals in endothelial cells by a pathway involving protein kinase C activation. Circ Res 71: 1422–1428

27. Kugiyama K, Kerns S, Morrisett JD, Roberts R, Henry PD (1990) Impairment of endothelium-dependent arterial relaxation by lysolecithin in modified low-density lipoproteins. Nature 344: 160–162

28. Kume N, Cybulsky MI, Gimbrone MA (1992) Lysophosphatidylcholine, a component of atherogenic lipoproteins, induces mononuclear leukocyte adhesion molecules in cultured human and rabbit arterial endothelial cells. J Clin Invest 90: 1138–1144

29. Latron Y, Chautan M, Anfosso F, Alessi MC, Nalbone G, Lafont H, Juhan-Vague I (1991) Stimulating effect of oxidized low density lipoproteins on plasminogen activator inhibitor-1 synthesis by endothelial cells. Arterisclerosis Thromb 11: 1821–1829

30. Libby P, Hansson GK (1991) Biology of disease. Involvement of the immune system in human atherogenesis: current knowledge and unanswered questions. Lab Invest 64: 5–15

31. Malden LT, Chait A, Raines EW, Ross R (1991) The influence of oxidatively modified low density lipoproteins on expression of platelet-derived growth factor by human monocyte-derived macrophages. J Biol Chem 266: 13901–13907

32. Mangin EL, Kugiyama K, Nguy JH, Kerns SA, Henry PD (1993) Effects of lysolipids and oxidatively modified low density lipoprotein on endothelium-dependent relaxation of rabbit aorta. Circ Res 72: 161–166

33. Masuda J, Ross R (1990) Atherogenesis during low level hypercholesterolemia in the nonhuman primate. I. Fatty streak formation. Arteriosclerosis 10: 164–177

34. Morel DW, DiCorleto PE, Chisolm GM (1984) Endothelial and smooth muscle cells alter low density lipoprotein in vitro by free radical oxidation. Arteriosclerosis 4: 357–364

35. Munro JM, van der Walt JD, Munro CS, Chalmers JAC, Cox EL (1987) An immunohistochemical analysis of human aortic fatty streaks. Hum Pathol 18: 375–380

36. Nègre-Salvayre A, Fitoussi G, Réaud V, Pieraggi M-T, Thiers J-C, Salvayre R (1992) A delayed and sustained rise of cytosolic calcium is elicited by oxidized LDL in cultured bovine aortic endothelial cells. FEBS 299: 60–65

37. Oishi K, Raynor RL, Charp PA, Kuo JF (1988) Regulation of protein kinase C by lysophospholipids. Potential role in signal transduction. J Biol Chem 263: 6865–6871

38. Palinski W, Ylä-Herttuala S, Rosenfeld ME, Butler SW, Socher SA, Parthasarathy S, Curtiss LK, Witztum JL (1990) Antisera and monoclonal antibodies specific for epitopes generated during oxidative modification of low density lipoprotein. Arteriosclerosis 10: 325–335

39. Parthasarathy S, Steinbrecher UP, Barnett J, Witztum JL, Steinber D (1985) Essential role of phospholipase A_2 activity in endothelial cell-induced modification of low density lipoprotein. Proc Natl Acad Sci USA 82: 3000–3004

40. Parthasarathy S, Barnett J (1990) Phospholipase A_2 activity of low density lipoprotein: evidence for an intrinsic phospholipase A_2 activity of apoprotein B-100. Proc Natl Acad Sci USA 87: 9741–9745

41. Prescott MF, Müller KR (1983) Endothelial regeneration in hypertensive and genetically hypercholesterolemic rats. Arteriosclerosis 3: 206–214
42. Quinn MT, Parthasarathy S, Steinberg D (1988) Lysophosphatidylcholine: a chemotactic factor for human monocytes and its potential role in atherogenesis. Proc Natl Acad Sci USA 85: 2805–2809
43. Resink TJ, Tkachuk VA, Bernhardt J, Bühler FR (1992) Oxidized low density lipoproteins stimulate phosphoinositide turnover in cultured vascular smooth muscle cells. Arteriosclerosis Thromb 12: 278–285
44. Rajavashisth TB, Andalibi A, Territo MC, Berliner JA, Navab M, Fogelman AM, Lusis AJ (1990) Induction of endothelial cell expression of granulocyte and macrophage colony-stimulating factors by modified low-density lipoproteins. Nature 344: 254–257
45. Rosenfeld ME, Palinski W, Ylä-Herttuala, Butler S, Witztum JL (1990) Distribution of oxidation specific lipid-protein adducts and apolipoprotein B in atherosclerotic lesions of varying severity from WHHL rabbits. Arteriosclerosis 10: 336–349
46. Schwartz CJ, Valente AJ, Sprague EA, Kelley JL, Nerem RM (1991) The patheogenesis of atherosclerosis: an overview. Clin Cardiol 14: I1 – I16
47. Shimokawa H, Flavahan NA, Vanhoutte PM (1991) Loss of endothelial pertussis toxin-sensitive G protein function in atherosclerotic porcine coronary arteries. Circulation 83: 652–660
48. Steinbrecher UP, Parthasarathy S, Leake DS, Witztum JL, Steinberg D (1984) Modification of low density lipoprotein by endothelial cells involves lipid peroxidation and degradation of low density lipoprotein phospholipids. Proc Natl Acad Sci USA 81: 3883–3887
49. Seuwen K, Pouysségur J (1992) G protein-controlled signal transduction pathways and the regulation of cell proliferation. Adv Can Res 58: 75–94
50. Walker LN, Bowyer DE (1984) Endothelial healing in the rabbit aorta and the effect of risk factors for atherosclerosis. Arteriosclerosis 4: 479–488
51. Weis JR, Pitas RE, Wilson BD, Rodgers GM (1991) Oxidized lowdensity lipoprotein increases cultured human endothelial cell tissue factor activity and reduces protein C activation. FASEB J 5: 2459–2465
52. Yamamoto H, Bossaller C, Cartwright J, Henry PD (1988) Videomicroscopic demonstration of defective cholinergic arteriolar vasodilation in atherosclerotic rabbit. J Clin Invest 81: 1752–1758
53. Yamamoto H, Sartori M, Cartwright J, Henry PD (1987) Carotid artery expansion after opposite carotid occlusion in rabbits: suppression by de-endothelization and hypercholesterolemia. Circulation 76: 55

Author's address:
Philip D. Henry, M.D.
Baylor College of Medicine
One Baylor Plaza, Suite 513E
Houston, Texas 77030
USA

Assessment of endothelial modulation of coronary vasomotor tone: Insights into a fundamental functional disturbance in vascular biology of atherosclerosis

A. M. Zeiher, V. Schächinger, B. Saurbier, H. Just

Department of Internal Medicine III, Division of Cardiology, University of Freiburg

Summary: The endothelium plays a major role in modulating vascular smooth muscle tone by synthesizing and metabolizing a number of vasoactive substances. Since the endothelium is both a target for and a mediator of vascular disease, functional alterations in coronary vascular reactivity due to endothelial dysfunction might play an important integral part in the clinical presentation of coronary artery disease. Recent advances in interventional techniques including intra-coronary instrumentation by Doppler catheters to measure blood flow velocities and 2-D-ultrasound catheters to evaluate arterial wall architecture during coronary angiography provided the diagnostic tools to assess endothelial vasodilator function and its relation to athero-sclerotic disease. The current weight of evidence suggests that disturbances of vasomotor function of epicardial conductance vessels are fundamental to the development of atherosclerosis, and impaired endothelial vasodilation is the predominant mechanism underlying inappropriate vasoconstriction in atherosclerosis. However, endothelial vasodilator dysfunction is not only confined to atherosclerotic epicardial vessels, but may also extend into the coronary microcirculation, which does not develop overt atherosclerotic lesions, but determines coronary blood flow in the absence of hemodynamically significant stenoses. The most important factors associated with impaired endothelium-mediated dilation of the coronary microcirculation are hypercholesterolemia and advanced age. With respect to the clinical presentation of coronary artery disease, endothelial vasodilator dysfunction appears to play a causative role for triggering myocardial ischemia in stable angina pectoris, to aggravate the sequelae of acute ischemic syndromes, and might be the primary underlying mechanism in some patients with syndrome X, whereas variant angina appears to be related to a hyperreactivity of the vascular smooth muscle layer.

Thus, the assessment of endothelium-mediated modulation of coronary vasomotor tone in the clinical setting offers unique and important insights into mechanisms leading to ischemic manifestations of coronary artery disease.

Key words: Endothelium − coronary artery disease − risk factors − acetylcholine − ultrasound − coronary blood flow

Coronary angiography has been used as a diagnostic tool to assess coronary atherosclerosis for more than 30 years, and decisions regarding specific interventions are still based upon the angiographic severity of stenosis. After all, however, coronary angiography is nothing more than a two-dimensional shadowgram of a contrast-filled vessel thereby merely demonstrating the effect of arterial wall disease on the contour of the arterial lumen. The underlying pathological process can be identified only by inference. However, ischemic manifestations of atherosclerosis including myocardial infarction are not necessarily associated with the preexistence of the most

"

severe stenosis, but often develop in the setting of mild-to-moderate angiographic stenoses (1, 44). Thus, it is not only the extent of a given atherosclerotic lesion, but rather the biological activity and possibly reactivity of each lesion, which appear to determine subsequent clinical events (16, 27).

Recent advances in interventional techniques including percutaneous intracoronary instrumentation with Doppler catheters to measure coronary blood flow velocities and 2-D ultrasound catheters to evaluate arterial wall architecture during cardiac catheterization provided the diagnostic tools to more closely examine one particular aspect of vessel wall function – the consequences of the disease process upon vascular reactivity of large and small coronary vessels in humans in vivo. These studies examining vasomotor function of the human coronary circulation during cardiac catheterization represent important advances to apply experimentally gained knowledge of the biology of the arterial wall to the clinical arena. This article will summarize features pertinent to epicardial artery and coronary resistance vessel reactivity in normal and diseased coronary arteries with special emphasis to endothelium-mediated mechanisms of regulation of vascular tone. The current weight of evidence suggests that disturbances in vasomotor function are fundamental to the development of atherosclerosis, and impaired endothelial vasodilation is the predominant mechanism underlying inappropriate vasoconstriction in atherosclerosis. Such functional alterations in coronary vascular reactivity might play an important integral part in the clinical presentation of coronary artery disease by affecting the balance between myocardial oxygen supply and demand. The assessment of endothelium-mediated modulation of coronary vasomotor tone in the clinical setting therefore offers unique and important insights into mechanisms leading to ischemic manifestations of coronary artery disease.

Endothelium-mediated modulation of vascular tone

The endothelium covers the inner surface of all blood vessels and has been recognized to play a major role in modulating vascular smooth muscle tone by synthesizing and metabolizing vasoactive substances (77) including an endothelium-derived relaxing factor (EDRF), an endothelium-derived hyperpolarizing factor, and prostacyclin (3, 25, 78). EDRF is released following the stimulation of muscarinic receptors on endothelial cells by acetylcholine as well as by other agonists or physical stimuli (26, 34). Produced by endothelial cells, this factor traverses the subendothelial space and activates smooth muscle cell guanylate cyclase to increase cyclic guanosine monophosphate levels leading to smooth muscle relaxation (38, 60). Evidence has been provided that nitric oxide (NO) – derived from L-arginine – or a related compound may account for the biological activity of EDRF (52, 56). NO/EDRF is continuously synthesized and released under basal conditions in most vascular beds and almost all species examined (2, 76, 77). The most important mechanism underlying basal NO/EDRF release is the shear exerted by the flowing blood upon the endothelial cell layer (11, 58, 65). The endothelial cells may be viewed as mechanotransducer sensing local blood flow and converting signals of increased shear stress into vessel wall relaxation thereby optimizing tissue perfusion according to the metabolic needs (33, 35). In normal physiological conditions, the rate of NO/EDRF production – and hence the caliber of the vessel – is directly related to the rate of flow. By virtue of the continuous adjustment of vessel caliber according to the prevailing blood flow the continuous basal release of NO/EDRF represents an important determinant of resting vascular resistance. Inhibition of endogenous NO/EDRF formation by the specific inhibitor L-monomethyl-arginine (L-NMMA) induces dramatic increases in systemic blood pressure in rats (61). Infusion of L-NMMA into the human forearm circulation led to a reduction in regional blood flow by approximately 50 % under resting baseline conditions (76). Thus, NO/EDRF does not only adjust the caliber of large conductance vessels, but – probably more importantly – also modulates the tone of resistance vessels, which determine regional blood flow and organ perfusion thereby coupling metabolic factors with resistance vessel tone.

In addition to the basal release of NO/EDRF mediated by the signal of shear stress, a number of endogenous substances such as acetylcholine, substance P, serotonin, thrombin, histamin, ADP, and ATP have been shown to stimulate endothelial NO/EDRF release via the activation of endothelial surface receptors (3, 26, 34, 77, 78). Some of these substances, like acetylcholine and serotonin, do not only stimulate the release of endothelial NO/EDRF, but at the same time exert direct constrictor effects via receptors on the vascular smooth muscle cells. Thus, the net effect of these substances reflects the balance between indirect endothelium-mediated vasodilation and direct vasoconstriction. In the pioneering experiments performed by Furchgott and Zawadzki (25) it could be demonstrated that in the presence of an intact endothelium acetylcholine produced dose-dependent relaxation of arterial segments, whereas removal of the endothelium resulted in contraction of similar arterial segments. These dichotomous responses to acetylcholine formed the basis to use the vasomotor response to acetylcholine as a diagnostic tool to assess the functional integrity of the endothelium. In addition, since acetylcholine does not normally circulate in vivo in the intact circulation and is rapidly inactivated by cholinesterases, it was regarded as an ideal parmacological test substance to assess endothelial vasodilator function in humans in vivo.

Endothelial function of epicardial arteries

Ludmer et al. (45) were the first to extend the experimental observations into the clinical setting by demonstrating that infusion of acetylcholine into the coronary system during diagnostic cardiac catheterization produced a dose-dependent vasodilation of normal epicardial vessels within a dose range of 10^{-9} to 10^{-6} M suggestive of an endothelium-dependent vasodilator mechanism to be operative in the human coronary circulation in vivo, too. Subsequent studies using either inactivators of NO/EDRF like methylene blue (37) or free hemoglobin (10) as well as the specific inhibitor L-NMMA (41) did provide indirect evidence that the vasodilator response of normal human epicardial arteries to acetylcholine is indeed – at least in part – mediated by the release of NO/EDRF, since inhibition of NO/EDRF effects converted the dilator responses of acetylcholine into a constrictor response. In addition, serotonin (30), histamine (48), and substance P (7, 15) were also shown to dilate normal epicardial arteries in vivo after intracoronary infusion during cardiac catheterization. Finally, our group (17) as well as Cox et al. (12) simultaneously reported that the mechanism of flow-dependent dilation is operative in normal human epicardial arteries and can be detected during coronary angiography. Thus, the diagnostic armamentarium to assess the functional integrity of the endothelium of epicardial arteries during cardiac catheterization comprises evaluation of both receptor-mediated (eg. acetylcholine) and receptor-independent (e.g. flow-mediated) mechanisms.

Since the endothelium is one of the primary targets of atherosclerosis and subendothelial atherosclerotic thickening of the intima may constitute an anatomical and/or functional barrier to the diffusion of NO/EDRF released from the endothelium (64), it was logical to propose that atherosclerosis will impair endothelium-mediated vasodilation. A large number of experimental studies indeed demonstrated defective endothelial vasodilation during diet-induced atherosclerosis in all species examined thus far (5, 23, 24, 36, 39). Similarly, in humans, the vasodilator response to acetylcholine observed in normal epicardial arteries was converted to a constrictor response in arteries with angiographic evidence of atherosclerosis (22, 45, 86) suggesting defective endothelium-mediated relaxation to counteract the direct constrictor effects upon the vascular smooth muscle cells. Subsequent clinical studies demonstrated that abnormal constrictor responses to acetylcholine were not only observed in angiographically diseased epicardial arteries, but also in angiographically normal appearing vessel segments of patients with evidence of atherosclerosis elsewhere in the coronary system and even in patients with entirely smooth epicardial arteries by angiography, but risk factors for coronary artery disease (79, 81, 84). The vasoconstrictor response to acetylcholine in angiographically normal epicardial arteries of patients with risk factors for coronary artery disease has been interpreted to reflect early atherosclerosis at a stage not detectable by angiography or even a disturbance of endothelial function that precedes the development of atherosclerosis (88). This interpretation was supported by studies demonstrating that arterial segments with the greatest propensity for the subsequent development of atherosclerosis – namely proximal segments and branching points – tended to constrict earlier in response to acetylcholine (51).

In addition, we could demonstrate that there is a progressive impairment in endothelium-mediated vasodilator function of epicardial arteries with different early stages of atherosclerosis as judged by angiography (88). Endothelium-dependent stimuli appear to be affected in a hierarchical fashion with receptor-dependent responses to be impaired before receptor-independent flow-mediated vasodilation. We recently corroborated these findings using intracoronary ultrasound examination to correlate arterial wall architecture with local vasomotor responses to different endothelial stimuli (91). These studies demonstrated that the acetylcholine response was converted into an abnormal constrictor response even in the presence of very minor atherosclerotic wall thickening as directly assessed in vivo by intracoronary ultrasound imaging, whereas flow-mediated dilation was well preserved despite ultrasonic evidence of atherosclerotic wall thickening and deteriorated only in vessel segments with advanced atherosclerotic wall abnormalities (92). Thus, abnormal vasomotor responses to acetylcholine do indeed provide the earliest evidence of endothelial vasoactive dysfunction in the human coronary circulation. These studies clearly illustrated the usefulness of intracoronary infusions of acetylcholine as a diagnostic test to detect abnormal local vascular reactivity during cardiac catheterization. In future prospective studies, it will be extremely important to examine whether a local defective endothelium-mediated dilator response of epicardial artery segments in patients with risk factors for coronary artery disease will ultimately be associated with the development of angiographically visible atherosclerotic lesions at a later stage of the disease process.

In parallel with studies assessing the coronary vasomotor responses to the endothelium-dependent dilator acetylcholine, a number of investigators demonstrated that epicardial artery segments exhibiting abnormal constrictor responses to acetylcholine also responded abnormally to a variety of common daily life stimuli like exercise (29, 31), mental stress (85), or cold exposure (53, 86). Whereas all these stimuli induced vasodilator responses in normal epicardial vessels, paradoxical constrictor responses were observed in atherosclerotic vessels. The mechanisms of vasodilation of normal epicardial coronary arteries in response to sympathetic activation induced by the above mentioned stimuli are believed to be mediated by endothelial α_2-adrenoceptor activation (59) and increased shear stress due to augmentation of coronary blood flow and driving pressure (87) thereby counteracting the direct α-receptor-mediated vascular smooth muscle constrictor effects (71). Thus, the endothelium appears to importantly modify the direct vasoconstrictor responses of catecholamines released during sympathetic activation. This hypothesis has been strongly supported very recently by studies of Vita et al. (80), who used the intracoronary infusion of phenylephrine to assess coronary vasoreactivity to catecholamines. These studies demonstrated that those epicardial artery segments, which exhibited a constrictor response to acetylcholine, also demonstrated a constrictor response to phenylephrine at a 100-fold lower concentration than segments with a dilator response to acetylcholine indicative of intact endothelial functioning. Thus, endothelial vasodilator dysfunction appears to render atherosclerotic vessels more sensitive to the constrictor effects of catecholamines. Taken together, these studies indicate that unopposed constriction due to a local failure of endothelium-mediated dilation causes atherosclerotic epicardial arteries to respond abnormally to sympathetic stimulation. Thus, endothelial vasodilator dysfunction might indeed play an important integral role in the pathogenesis of inappropriate vasoconstriction, which represents the fundamental functional disturbance in vascular biology of atherosclerosis.

Endothelial function of coronary resistance vessels

In the absence of hemodynamically significant epicardial artery atherosclerotic lesions, abnormal vasoreactivity of conduit vessel function probably contributes little to the regulation of coronary blood flow, because myocardial perfusion is regulated predominantly by microvessels $< 200\ \mu$m in diameter (8). However, unlike the walls of larger arteries, the walls of resistance vessels do not develop overt atheroma (40). Nevertheless, several recent experimental studies did show that endothelium-dependent relaxation is abnormal in the resistance vessels of cholesterol-fed atherosclerotic animals (54, 68, 82). These results suggested that, despite the heterogeneity in the vascular susceptibility to develop overt atheroma in response to the atherogenic substance, the exposure of blood vessels to increased levels of circulating cholesterol induces abnormalities in vascular function in both conductance and resistance vessels. Especially early in the course of coronary atherosclerotic disease prior to the development of hemodynamically significant epicardial artery lesions, an altered endothelial regulation of blood vessels would be of substantial pathophysiological importance if it would occur at flow-regulating sites within the coronary circulation.

Assessment of coronary resistance vessel function requires the measurement of coronary blood flow during cardiac catheterization. Combining epicardial artery diameter measurements obtained by quantitative angiography with intracoronary blood flow velocities subselectively measured by Doppler catheters allows at least for an estimate of coronary blood flow and its alterations during pharmacological interventions. However, it should be kept in mind that these methods do provide no direct data to identify the specific site of an observed impaired dilator response of the coronary microvessels.

Regardless of these methodological limitations, we and others (74, 88, 93) could demonstrate that acetylcholine profoundly increases coronary blood flow in normal humans by approximately 250 % corresponding to a decrease in coronary vascular resistance by more than 70 %. The endothelium-mediated increase in coronary blood flow amounted to roughly one half of the maximally achievable increase in blood flow by the smooth muscle relaxant papaverine (93). Thus, the stimulated release of NO/EDRF activity by acetylcholine can substantially modify coronary vascular resistance in the intact human coronary circulation in vivo. Importantly, failure of endothelium-mediated dilation within the coronary resistance vessels may aggravate the functional consequences of atherosclerosis by limiting increases in blood flow during increased demand, e.g. during exercise or sympathetic activation.

Indeed, we could demonstrate in an earlier study that in some patients with evidence of early epicardial atherosclerosis endothelium-mediated dilation of the coronary resistance vasculature was impaired, indicating that the pathophysiological consequences of atherosclerosis may extend into the human coronary microcirculation (89). Importantly, impaired endothelium-mediated relaxation of the coronary microvasculature was closely correlated with the failure of coronary blood flow to increase during cold exposure suggesting that endothelial dysfunction of the coronary microvessels uncouples resistance vascular tone from metabolic factors (89). These results have been recently confirmed by experimental studies demonstrating that endothelium-mediated resistance vessel relaxation protects against an adrenoceptor-mediated increases in coronary vascular resistance during sympathetic activation (46). Thus, the functional integrity of the endothelium seems to be crucial for the regulation of coronary blood flow during increased metabolic demand.

However, since not all patients with evidence of atherosclerosis in their epicardial arteries do exhibit a reduced dilator response of their resistance vessels to acetylcholine, additional factors must be operative, which may independently affect endothelial function of the coronary microvasculature and thereby exacerbate the pathophysiological consequences of coronary artery disease. Irrespective of epicardial artery reactivity, the most important factors associated with an impaired coronary blood flow response to acetylcholine are elevated serum cholesterol levels and advanced age (93).

We have previously shown that the coronary blood flow response to acetylcholine was considerably impaired in patients with hypercholesterolemia compared to patients with normal cholesterol levels despite similar reductions in epicardial artery cross-sectional areas and similar papaverine-induced increases in coronary blood flow (88). Similar results have been recently published by Egashira et al. (21). The blunted blood flow response to acetylcholine appears to be directly related to the total serum cholesterol level (21, 93). Interestingly, a blunted acetylcholine-mediated increase in blood flow has also been observed in the human forearm circulation (13). Thus, elevated cholesterol levels appear to be associated with a generalized endothelial dysfunction of human resistance vessels.

The mechanisms responsible for the blunted acetylcholine-induced relaxation of human microvessels in hypercholesterolemia remain to be determined. Importantly,

120

both the coronary microvessels as well as the forearm resistance vessels do not develop grossly visible atherosclerotic lesions (40, 55), although electron microscopy demonstrated the presence of vacuoles likely representing lipid droplets within the endotelium of microvessels obtained from diet-induced atherosclerotic animals (68). An interference of lipids with receptor-operated signal transduction mechanisms linked to the formation of EDRF, specifically with the receptor-mediated intracellular availability of L-arginine (70), might be a potential mechanism for decreased NO/EDRF activity in hypercholesterolemia. Indeed, the supplementation of L-arginine has been shown by our group (18) and by Creager et al. (14) to acutely normalize endothelium-dependent responses in resistance vessels of hypercholesterolemic, but not of normocholesterolemic subjects. Thus, an interference of lipoproteins with the receptor-mediated intracellular availability of L-arginine, the precursor of EDRF, might indeed play an important role for the blunted acetylcholine-induced increase in coronary blood flow observed in patients with elevated serum cholesterol levels.

The mechanisms responsible for the age-associated reduction in the ability of the coronary microvasculature to dilate in response to the pharmacological stimulation with acetylcholine remain to be determined, too. Experimental studies demonstrating thinning and loss of endothelial cells in aged animals (69) suggested an impaired production of EDRF (49). In addition, advanced glycosylation endproducts, which accumulate in the vascular subendothelium in aging (67), have been experimentally shown to quench nitric oxide and mediate defective endothelium-dependent vasodilation (6). Other possible mechanisms include a decline in endothelial muscarinic receptor density, the release of hyperpolarizing or constricting factors, or an increased sensitivity of vascular smooth muscle to the constrictor effect of acetylcholine with advancing age. Interestingly, Nabel et al. (53) observed a significantly impaired coronary blood flow response to increases in heart rate obtained by cardiac pacing in elderly patients suggesting that advanced age is also associated with an increased constrictor response during sympythetic stimulation.

Surprisingly, hypertension per se does not appear to be associated with endothelial dysfunction of the coronary resistance vasculature (93). Patients with hypertension and left ventricular hypertrophy frequently have signs and symptoms of myocardial ischemia in the absence of obstructive epicardial artery disease indicative of a potential impairment of microvascular relaxation. Two recently published studies suggested that hypertension is associated with impaired endothelium-mediated relaxation in human coronary resistance vessels (21, 75). However, most of these patients had additional risk factors like hypercholesterolemia, which are know to impair endothelium-mediated increases in coronary blood flow. Consequently, when a multivariate analysis was used to take into account multiple factors implicated in defective endothelium-dependent dilation, a history of hypertension was no longer significantly associated with an impaired relaxation of coronary resistance vessels (21, 75). Indeed, we could not demonstrate a blunted coronary blood flow response to acetylcholine in an age-matched normocholesterolemic group of patients with a history of hypertension compared to normocholesterolemic patients with normotensive blood pressures (93). Thus, endothelial dysfunction in hypertensive patients appears to be confined to the large epicardial vessels, which are continuously exposed to high pulsatile pressure and shear stress (19). These findings contrast with the results of studies in the human forearm circulation, where arterial hypertension has been shown to be associated with a blunted blood flow response to acetylcholine despite normal responses to nitroprusside (42, 57). These discrepancies might be simply related to the different vascular beds. Whereas the epicardial arteries – like the large arteries of the cerebral and limb circulation – are known targets of hypertension (62), hypertensive vascular disease rarely develops in the large vessels of the human forearm circulation.

Endothelial dysfunction and clinical symptoms of coronary artery disease

Endothelium-mediated relaxation is impaired in atherosclerotic human coronary arteries shifting the balance in favor of vasoconstriction in response to a variety of stimuli, most notably in response to sympathetic activation. In addition, atherosclerosis impairs flow-mediated dilation of epicardial conductance vessels, and endothelium-mediated vasodilator function of the coronary resistance vessels is adversely affected by hypercholesterolemia and advanced age. Although none of these functional alterations by itself may reduce coronary blood flow below myocardial oxygen demands, defective endothelium-mediated vasodilator functions may potentiate known trigger mechanisms of myocardial ischemia and thereby induce a mismatch between myocardial oxygen supply and demand.

Sympathetic activation is an important trigger mechanism of myocardial ischemia in patients with stable angina. Exercise, mental stress, and cold exposure are welldocumented stimuli to induce myocardial ischemia in patients with stable angina pectoris. Clearly, inappropriate vasoconstriction due to endothelial vasodilator dysfunction of atherosclerotic epicardial arteries might play a pathogenetic role in the precipitation of myocardial ischemia in patients with epicardial artery stenoses. Even though the changes in luminal diameter produced by inappropriate vasoconstriction in response to sympathetic activation are usually less than 30 %, such an increase in arterial tone might be enough to convert a subcritical into a critical stenosis with ensuing decreases in blood flow. The resistance to flow offered by a stenosis is expected to quadruple as the degree of diameter narrowing increases from 70 % to 90 % (32) with resting blood flow remaining constant unless stenosis severity exceeds 80 % diameter narrowing. Thus, for stenoses with > 60 % diameter narrowing the abnormal vasoconstrictor response due to defective endothelial dilation may indeed be responsible for the mismatch between myocardial oxygen supply and demand indicative of a pathogenetic link between endothelial vasodilator dysfunction at the site of the stenosis and myocardial ischemia associated with routine daily life activities. In addition, impaired flow-mediated dilation of epicardial vessels and reduced endothelium-mediated dilator capacity of coronary resistance vessels will further reduce coronary flow reserve and therefore contribute to a further reduction in the ischemic threshold during increased myocardial demands. Thus, the functional integrity of both epicardial and coronary resistance vascular endothelium will determine the balance between myocardial oxygen supply and demand during sympathetic activation associated with increased myocardial work in patients with stable angina.

In stable coronary artery disease, there is a circadian variation in ischemic events, with transient ischemic episodes being most frequent in the morning hours (63). In addition to increases in heart rate and blood pressure, humoral vasoconstrictors like norepinephrine (43) and platelet aggregability (73) are increased during the morning hours compared to other times of the day. Indeed, vascular resistance has been shown to be elevated and ischemic threshold to be reduced in the morning hours (59).

Given the important role of the endothelium to modulate vasomotor responses to sympathetic activation and to vasoactive substances released by aggregating platelets, deficient endothelium-dependent dilation may very likely contribute to circadian variation in ischemic episodes in stable coronary artery disease by facilitating increases in coronary vascular resistance due to inappropriate vasoconstrictor responses.

In contrast, the pathomorphological correlate of unstable angina is characterized by plaque fracture and ensuing platelet activation and thrombus formation (28). It is hardly imaginable that endothelial vasodilator dysfunction does play a causative role for the occurrence of plaque fracture. However, when platelet deposition and thrombus formation follows the initial event of plaque fracture, a number of vasoactive substances are directly released into the coronary circulation, most notably serotonin and thrombin. Both substances have been shown to exert potent vasoconstrictor effects in the presence of a dysfunctional endothelium, whereas normal endothelial function mediates a dilator response (3, 25, 77, 78). Thus, endothelial dysfunction may importantly magnify the constrictor responses to substances released during platelet activation and thrombus formation not only at the site of plaque rupture, but most importantly in the distal vascular bed and especially in the coronary resistance vasculature thereby impairing coronary blood flow regulation. Indeed, we have demonstrated that intracoronary thrombus formation causes profound vasoconstriction distal to the site of the thrombus in patients with atherosclerosis (90). Moreover, two recent studies demonstrated that the intracoronary infusion of serotonin induces dramatic increases in coronary vascular resistance indicative of impaired dilator responses of the coronary microcirculation in patients with evidence of atherosclerosis (30, 50). Experimental studies have demonstrated that diet-induced hypercholesterolemia considerably increases infarct size as well as ischemia-reperfusion injury in experimental animals (54, 66, 72). In light of recent experimental evidence that the endothelium of the coronary microcirculation is extraordinarily sensitive to ischemia, responding with loss of natural anticoagulant and vasorelaxing properties (4), it is very intriguing to hypothesize that the blunted endothelium-mediated dilator capacity of the coronary microvasculature may play an important pathophysiological role for the aggravated sequelae of acute ischemic events in elderly and hypercholesterolemic patients with coronary artery disease.

The trigger mechanisms of frank coronary spasm, which is characteristic of the clinical syndrome of variant angina, are still poorly understood. However, current evidence suggests that it is rather related to a hypersensitivity of the vascular smooth muscle cells (47) than the consequence of impaired endothelium-mediated dilation. This hypothesis is also supported by studies demonstrating a normal endothelial dilator effect to substance P in coronary artery segments demonstrating hyperreactivity to ergonovine in patients with variant angina (20, 83).

In summary, advances in interventional catheter technology and intracoronary instrumentation have enabled the comprehensive functional assessment of endothelium-mediated coronary vasomotor tone regulation in patients during cardiac catheterization. Thus, knowledge gained from basic vascular biology can now be applied in the clinical setting. These studies extended experimental data regarding the fundamental importance of a functionally intact endothelium for the circulatory homeostasis into the human coronary circulation. It is now clear that atherosclerosis severely impairs endothelial vasodilator function not only of epicardial conductance vessels, but also – in part – of the coronary resistance vasculature, which regulates myocardial perfusion and determines the balance between oxygen demand and supply. Convincing evidence has been provided that a dysfunctional endothelium alters the balance of humoral and neural factors in favor of vasoconstrictor influences, which facilitate the manifestation of myocardial ischemia. Thus, these studies offered unique and important insights into a fundamental disturbance in vascular biology of atherosclerosis, namely inappropriate vasoconstriction. Future studies addressing the molecular mechanisms underlying endothelial vasodilator dysfunction and most notably its potential reversibility by therapeutic strategies will undoubtedly be of benefit for the treatment of myocardial ischemia in the short term and of coronary atherosclerosis in the long term.

References

1. Ambrose JA, Tannenbaum MA, Alexopoulos et al. (1988) Angiographic progression of coronary artery disease and the development of myocardial infarction. J Am Coll Cardiol 12: 56–62
2. Amezcua JL, Palmer RMJ, de Souza BM, Moncada S (1989) Nitric oxide synthesized from L-arginine regulates vascular tone in the coronary circulation of the rabbit. Br J Pharmacol 97: 1119–1124
3. Bassenge E, Busse R (1980) Endothelial modulation of coronary tone. Progr Cardiovasc Dis 30: 349–380
4. Berk BC (1991) The microcirculation in coronary ischemia. Circulation 84: 439–441
5. Bossaler C, Habib GB, Yamamoto H, Williams C, Wells S, Henry PD (1987) Impaired muscarinic endothelium-dependent relaxation and cyclic guanosine 5-monophosphate formation in atherosclerotic human coronary artery and rabbit aorta. J Clin Invest 79: 170–174
6. Bucala R, Tracey KJ, Cerami A (1991) Advanced glycosylation products quench nitric oxide and mediate defective endothelium-dependent vasodilation in experimental diabetes. J Clin Invest 87: 432–438
7. Chester AH, O'Neil GS, Moncada S, Tadjkarimi S, Yacoub M (1990) Low basal and stimulated release of nitric oxide in atherosclerotic epicardial coronary arteries. Lancet 336: 897–900
8. Chilian WM, Eastham CL, Marcus ML (1986) Microvascular distribution of coronary vascular resistance in beating left ventricle. Am J Physiol 251: H779–H786
9. Cocks TM, Angus JA (1983) Endothelium-dependent relaxation in canine large arteries. Nature 305: 627–630
10. Collins P, Burman J, Chung H-I, Fox K (1993) Hemoglobin inhibits endothelium-dependent relaxation to acetylcholine in human coronary arteries in vivo. Circulation 87: 80–85
11. Cooke JP, Rossitch E Jr, Andon NA, Loscalzo J, Dzau VJ (1991) Flow activates an endothelial potassium channel to release an endogenous nitrovasodilator. J Clin Invest 88: 1663–1671
12. Cox DA, Vita JA, Treasure CB et al. (1989) Impairment of flow-mediated coronary dilation by atherosclerosis in man. Circulation 80: 458–465
13. Creager MA, Cooke JP, Mendelsohn ME, Gallagher SJ, Coleman SM, Loscalzo J, Dzau VJ (1990) Impaired vasodilation of forearm resistance vessels in hypercholesterolemic humans. J Clin Invest 86: 228–234
14. Creager MA, Gallagher SJ, Girerd XJ et al. (1992) L-arginine improves endothelium-dependent vasodilation in hypercholesterolemic humans. J Clin Invest 90: 1248–1253
15. Crossman DC, Larkin SW, Fuller RW, Davies GJ, Maseri A (1989) Substance P dilates epicardial coronary arteries and increases coronary blood flow in humans. Circulation 80: 475–484

16. Davies MJ,Thomas AC (1985) Plaque fissuring – the cause of acute myocardial infarction, sudden ischemic death, and crescendo angina. Br Heart J 53: 363–373
17. Drexler H, Zeiher AM, Wollschläger H, Meinertz T, Just H, Bonzel T (1989) Flow-dependent coronary artery dilatation in humans. Circulation 80: 466–474
18. Drexler H, Zeiher AM, Meinzer K, Just H (1991) Correction of endothelial dysfunction in coronary microcirculation of hypercholesterolemic patients by L-arginine. Lancet 338: 1546–1550
19. Dzau VJ, Safar ME (1988) Large conduit arteries in hypertension: role of the vascular renin-angiotensin system. Circulation 77: 947–954
20. Egashira K, Inou T, Yamada A, Hirooka Y, Takeshita A (1992) Preserved endothelium-dependent vasodilation at the vasospastic site in patients with variant angina. J Clin Invest 89: 1047–1052
21. Egashira K, Inou T, Hirooka Y et al. (1993) Impaired coronary blood flow response to acetylcholine in patients with coronary risk factors and proximal atherosclerotic lesions. J Clin Invest 91: 29–37
22. Fish RD, Nabel EG, Selwyn AP et al. (1988) Responses of coronary arteries of cardiac transplant patients to acetylcholine. J Clin Invest 81: 21–31
23. Förstermann U, Mügge A, Alheid U, Haverich A, Frölich J (1988) Selective attenuation of endothelium-mediated vasodilation in atherosclerotic human coronary arteries. Circ Res 62: 185–190
24. Freiman PC, Mitchell GC, Heistad DD, Armstrong ML, Harrison DG (1986) Atherosclerosis impairs endothelium-dependent vascular relaxation to acetylcholine and thrombin in primates. Circ Res 58: 783–789
25. Furchgott RF, Zawadzki JV (1980) The obligatory role of endothelial cells in the relaxation of arterial smooth muscle by acetylcholine. Nature 288: 373–376
26. Furchgott RF (1983) The role of endothelium in response of vascular smooth muscle. Circ Res 53: 557–573
27. Fuster V, Stein B, Ambrose JA, Badimon L, Badimon JJ, Chesebro JH (1990) Atherosclerotic plaque rupture and thrombosis: evolving concepts. Circulation 82 (suppl II): II-47–II-59
28. Fuster V, Badimon L, Badimon J, Chesebro J (1992) Mechanisms of disease: the pathogenesis of coronary artery disease and the acute coronary syndromes I. N Engl J Med 326: 242–250
29. Gage JE, Hess OM, Murakami T, Ritter M, Grimm J, Krayenbuehl HP (1986) Vasoconstriction of stenotic coronary arteries during dynamic exercise in patients with classic angina pectoris: reversibility by nitroglycerin. Circulation 73: 865–76
30. Golino P, Piscione F, Willerson JT, Cappelli-Bigazzi M, Focaccio A, Villari B, Indolfi C, Russolillo E, Condorelli M, Chiariello M (1991) Divergent effect of serotonin on coronary-artery dimensions and blood flow in patients with coronary atherosclerosis and control patients. N Engl J Med 324: 641–648
31. Gordon JB, Ganz P, Nabel EG et al. (1989) Atherosclerosis influences the vasomotor response of epicardial coronary arteries to exercise. J Clin Invest 83: 1946–52
32. Gould KL (1985) Quantification of coronary artery stenosis in vivo. Circ res 57: 341–353
33. Griffith TM, Edwards DH, Davies RL, Harrison TJ, Evans KT (1987) EDRF coordinates the behaviour of vascular resistance vessels. Nature 329: 442–445
34. Griffith TM, Lewis MJ, Newby AC, Henderson AH (1988) Endothelium-derived relaxing factor. J Am Coll Cardiol 12: 797–806
35. Griffith TM, Edwards DH (1990) Myogenic autoregulation of flow may be inversely related to endothelium-derived relaxing factor activity. Am J Physiol 258: H1171–H1180
36. Habib JBN, Bossaller C, Wells S, Williams C, Morrisett JD, Henry PD (1986) Preservation of endothelium-dependent vascular relaxation in cholesterol-fed rabbit by treatment with the calcium blocker PN 200110. Circ Res 58: 305–309
37. Hodgson JMcB, Marshall J (1989) Direct vasoconstriction and endothelium-dependent vasodilation. Mechanisms of acetylcholine effects on coronary flow and arterial diameter in patients with nonstenotic coronary arteries.Circulation 79: 1043–1051

38. Ignarro LJ, Harbison RG, Wood KS, Kadowitz PJ (1986) Activation of purified soluble guanylate cyclase by endothelium-derived relaxing factor from intrapulmonary artery and vein: stimulation by acetylcholine, bradykinin and arachidonic acid. J Pharmacol Exp Ther 237: 893–900

39. Jayakody L, Senaratne M, Thomson A, Kappagoda T (1987) Endothelium-dependent relaxation in experimental atherosclerosis in the rabbit. Circ Res 60: 251–264

40. Juergens JL, Bernatz PE (1980) Atherosclerosis of the extremities. In: Peripheral Vascular Diseases. Juergens JL, Spittell JA, Fairbairn JF II, editors. WB Saunders, Company, Philadelphia, PA: 253–293

41. Lefroy DC, Crake T, Uren NG, Davies GJ, Maseri A (1992) Affect of nitric oxide in the human coronary circulation (abstract). Circulation 86 (suppl I): I–118

42. Linder L, Kiowski W, Bühler FR, Lüscher T (1990) Indirect evidence for release of endothelium-derived relaxing factor in human forearm circulation in vivo. Blunted response in essential hypertension. Circulation 81: 1762–1767

43. Linsell CR, Lightman SL, Mullen PE, Brown MJ, Causon RC (1985) Circadian rhythms in epinephrine and norepinephrine in man. J Clin Endocrinol Metab 60: 1210–1215

44. Little WC, Constaninescu M, Applegate RJ et al. (1988) Can coronary angiography predict the site of a subsequent myocardial infarction in patients with mild-to-moderate coronary artery disease? Circulation 78: 1157–1166

45. Ludmer PL, Selwyn AP, Shook TL, Wayne RR, Mudge GH, Alexander RW, Ganz P (1986) Paradoxical vasoconstriction induced by acetylcholine in atherosclerotic coronary arteries. N Engl J Med 315: 1046–1051

46. Marcus ML, Chilian WM, Kanatsuka H, Dellsperger KC, Eastham CL, Lamping KG (1990) Understanding the coronary circulation through studies at the microvascular level. Circulation 82: 1–7

47. Maseri A, Davies G, Hackett D, Kaski JC (1990) Coronary artery spasm and vasoconstriction. The case for a distinction. Circulation 81: 1983–1991

48. Matsuyama K, Yasue H, Okumura K et al. (1990) Effects of H_1-receptor stimulation on coronary arteriel diameter and coronary hemodynamics in humans. Circulation 81: 65–71

49. Mayhan WG, Faraci FM, Baumbach GL, Heistad DD (1990) Effects of aging on responses of cerebral arterioles. Am J Physiol 258: H1138–H1143

50. McFadden EP, Clarke JG, Davies GJ, Kaski JC, Haider AW, Maseri A (1991) Effect of intra-coronary serotonin on coronary vessels in patients with stable angina and patients with variant angina. N Engl J Med 324: 648–654

51. McLenachan JM, Vita JA, Fish RD et al. (1990) Early evidence of endothelial vasodilator dysfunction at coronary branchpoints. Circulation 82: 1169–1173

52. Myers PR, Minor RL, Guerra R, Bates JN, Harrison DG (1990) Vasorelaxant properties of the endothelium-derived relaxing factor more closely resemble S-nitrosocysteine than nitric oxide. Nature 345: 161–3

53. Nabel EG, Ganz P, Gordon JB, Alexander RW, Selwyn AP (1988) Dilation of normal and constriction of atherosclerotic coronary arteries caused by the cold pressor test. Circulation 77: 43–52

54. Osborne JA, Lento PH, Siegfried MR, Stahl GL, Fusman B, Lefer AM (1989) Cardiovascular effects of acute hypercholesterolemia in rabbits: Reversal with lovastatin treatment. J Clin Invest 83: 465–473

55. Osborne JA, Siegman MJ, Sedar AW, Mooers SU, Lefer AM (1989) Lack of endothelium-dependent relaxation in coronary resistance arteries of cholesterol-fed rabbits. Am J Physiol 256: C591–C597

56. Palmer RM, Ferrige AG, Moncada S (1987) Nitric oxide accounts for the biological activity of endothelium-derived relaxing factor. Nature 327: 524–526

57. Panza JA, Quyyumi AA, Brush JE, Epstein SE (1990) Abnormal endothelium-dependent vascular relaxation in patients with essential hypertension. N Engl J Med 323: 22–27

58. Pohl U, Holtz J, Busse R, Bassenge E (1986) Crucial role of endothelium in the vasodilator response to increased flow in vivo. Hypertension 8: 37–44

59. Quyyumi AA, Panza JA, Diodati JG, Lakatos E, Epstein SE (1992) Circadian variation in ischemic threshold. A mechanism underlying the circadian variation in ischemic events. Circulation 86: 22–28

60. Rapoport RM, Murad F (1983) Agonist-induced endothelium-dependent relaxation in rat thoracic aorta may be mediated through cGMP. Circ Res 52: 352–357

61. Rees DD, Palmer RM, Moncada S (1989) Role of endothelium-derived nitric oxide in the regulation of blood pressure. Proc Natl Acad Sci USA 86: 3375–3378

62. Roberts WC (1980) The hypertensive diseases. In: Topics in Hypertension. Laragh JH, editor. Yorke Medical Books, New York: 368–388

63. Rocco MB, Barry J, Campbell S et al. (1989) Circadian variation of transient myocardial ischemia in patients with coronary artery disease. Circulation 75: 395–400

64. Ross R (1986) The pathogenesis of atherosclerosis – An update. N Engl J Med 314: 488–500

65. Rubanyi GM, Romero JC, Vanhoutte PM (1986) Flow-induced release of endothelium-derived relaxing factor. Am J Physiol 250: H1145–H1149

66. Sakamoto S, Kashiki M, Imai N, Liang C, Hood WB Jr. (1991) Effects of short-term, diet-induced hypercholesterolemia on systemic hemodynamics, myocardial blood flow, and infarct size in awake dogs with acute myocardial infarction. Circulation 84: 378–386

67. Schnider SL, Kohn RR (1980) Glucosylation of human collagen in aging and diabetes mellitus. J Clin Invest 66: 1179–1181

68. Sellke FW, Armstrong ML, Harrison DG (1990) Endothelium-dependent vascular relaxation is abnormal in the coronary microcirculation of atherosclerotic primates. Circulation 81: 1586–1593

69. Stewart PA, Magliocco M, Hayakawa K, Farrell CL, Del Maestro FR, Girvin J, Kaufmann JCE, Vinters HV, Gilbert J (1987) A quantitative analysis of blood-brain barrier ultrastructure in the aging human. Microvasc Res 33: 270–282

70. Tanner FC, Noll G, Boulanger CM, Lüscher TF (1991) Oxidized low density lipoproteins inhibit relaxations of porcine coronary arteries. Role of scavenger receptor and endothelium-derived nitric oxide. Circulation 83: 2012–2020

71. Tesfamarian B, Cohen RA (1988) Inhibition of adrenergic vasoconstriction by endothelial cell shear stress. Circ Res 63: 720–725

72. Tilton RP, Cole PA, Zions JD, Daugherty A, Larson KB, Sutera SP, Kilo C, Williamson JR (1987) Increased ischemia-reperfusion injury to the heart associated with short-term, diet-induced hypercholesterolemia in rabbits. Circ Res 60: 551–559

73. Tofler GH, Brezinski D, Schafer Al et al. (1987) Concurrent morning increase in platelet aggregability and the risk of myocardial infarction and sudden cardiac death. N Engl J Med 316: 1514–1518

74. Treasure CB, Vita JA, Cox DA, Fish R, Gordon JB, Mudge GH, Colucci WS, St John Sutton MG, Selwyn AP, Alexander RW, Ganz P (1990) Endothelium-dependent dilation of the coronary microvasculature is impaired in dilated cardiomyopathy. Circulation 81: 772–779

75. Treasure CB, Klein JL, Vita JA et al. (1993) Hypertension and left ventricular hypertrophy are associated with impaired endothelium-mediated relaxation in human coronary resistance vessels. Circulation 87: 86–93

76. Vallance P, Collier J, Moncada S (1989) Effects of endothelium-derived nitric oxide on peripheral arteriolar tone in man. Lancet 2: 997–1000

77. Vane R, Anggard EE, Botting RM (1990) Regulatory functions of the vascular endothelium. N Engl J Med 323: 27–36

78. Vanhoutte PM, Rubanyi GM, Miller VM, Houston DS (1986) Modulation of vascular smooth muscle concentration by the endothelium. Ann Rev Physiol 48: 307–320

79. Vita JA, Treasure CB, Nabel EG, McLenachan JM, Fish RD, Yeung AC, Vekshtein VI, Selwyn AP, Ganz P (1990) Coronary vasomotor response to acetylcholine relates to risk factors for coronary artery disease. Circulation 81: 491–497

80. Vita JA, Treasure CB, Yeung AC et al. (1992) Patients with evidence of coronary endothelial dysfunction as assessed by acetylcholine infusion demonstrate marked increase in sensitivity to constrictor effects of catecholamines. Circulation 85: 1390–1397

81. Werns SW, Walton JA, Hsia HH, Nabel EG, Sanz ML, Pitt B (1989) Evidence of endothelial dysfunction in angiographically normal coronary arteries of patients with coronary artery disease. Circulation 79: 287–291
82. Yamamoto H, Bossaller C, Cartwright J, Henry PD (1988) Videomicroscopic demonstration of defective cholinergic arteriolar vasodilation in atherosclerotic rabbit. J Clin Invest 81: 1752–1758
83. Yamamoto H, Yoshimura H, Noma M et al. (1990) Evaluation of endothelial function of spastic segments in patients with variant angina. Circulation 82: (suppl II): II-421 (abstract)
84. Yasue H, Matsuyama K, Matsuyama K, Okumura K, Morikami Y, Ogawa H (1990) Responses of angiographically normal human coronary arteries to intracoronary injection of acetylcholine by age and segment. Circulation 81: 482–490
85. Yeung AC, Vekshtein VI, Krantz DS et al. (1991) The effect of atherosclerosis on the vasomotor response of coronary arteries to mental stress. N Engl J Med 325: 1551–6
86. Zeiher AM, Drexler H, Wollschläger H, Saurbier B, Just H (1989) Coronary vasomotion in response to sympathetic stimulation in humans: importance of the functional integrity of the endothelium. J Am Coll Cardiol 14: 1181–1190
87. Zeiher AM, Drexler H (1990) Coronary hemodynamic determinants of epicardial artery vasomotor responses during sympathetic stimulation in humans. Bas Res Cardiol 86 (suppl II): 203–213
88. Zeiher AM, Drexler H, Wollschläger H, Just H (1991) Modulation of coronary vasomotor tone. Progressive endothelial dysfunction with different early stages of coronary atherosclerosis. Circulation 83: 391–401
89. Zeiher AM, Drexler H, Wollschläger H, Just H (1991) Endothelial dysfunction of the coronary microvasculature is associated with impaired coronary blood flow regulation in patients with early atherosclerosis. Circulation 84: 1984–1992
90. Zeiher AM, Schächinger V, Weitzel SH, Wollschläger H, Just H (1991) Intracoronary thrombus formation causes focal vasoconstriction of epicardial arteries in patients with coronary artery disease. Circulation 83: 1519–1525
91. Zeiher AM, Hohnloser SH, Fritz R et al. (1992) Acetylcholine-induced constrictor responses of angiographically normal coronary artery segments invariably predict the presence of atherosclerotic lesions. J Am Coll Cardiol 19: 242A (abstract)
92. Zeiher AM, Saurbier B, Hohnloser SH, Schächinger V, Bleile T, Fritz R (1992) Epicardial artery vasoreactivity and wall architecture in patients with early atherosclerosis. Circulation 86: I-117 (abstract)
93. Zeiher AM, Drexler H, Saurbier B, Just H (1993) Endothelium-mediated coronary blood flow modulation in humans: Effects of age, atherosclerosis, hypercholesterolemia, and hypertension. J Clin Invest 92: 652–662

Authors' address:
A. M. Zeiher, M.D.
Department of Internal Medicine III
Division of Cardiology
University of Freiburg
Hugstetterstr. 55
D-79106 Freiburg
FRG

In vitro assessment of luminal dimensions of coronary arteries by intravascular ultrasound with and without application of echogenic contrast dye

K. M. Schmid, W. Voelker, J. Mewald, H.-J. Paul[2], M. Wehrmann[2], B. Bültmann[2], K. R. Karsch

Departments of Cardiology and [2]Pathology, University of Tübingen, Tübingen, FRG

Summary: To evaluate the impact and limitations of intracoronary ultrasound in the assessment of lumen, we examined 80 segments of 20 isolated coronary arteries with a mechanical ultrasound device (CVIS) comparing the results of ultrasound with the corresponding histological specimens. Ultrasound was performed with and without echogenic contrast dye (Laevovist, Schering AG, FRG).

After application of contrast dye, correlation of luminal area between histology and ultrasound was improved from r = 0.85 to r = 0.89 (p = ns). Accuracy of lumen measurements was low in segments <2.5 mm; only after application of contrast dye a relationship between ultrasound and histological measurements was found.

In all cases in which a deviation of more than 20 % between ultrasound and histology was observed, this deviation could be reduced by the application of contrast dye.

There are considerable limitations in the accuracy of ultrasound measurements in the near field. Thus, further improvement of intracoronary ultrasound devices is mandatory. However, with the use of the currently available systems, additional application of echogenic contrast dye can improve accuracy of luminal measurements, especially in smaller size vessels.

Key words: Intravascular ultrasound – echogenic contrast dye – coronary artery disease

Introduction

Catheter-based, coronary imaging by intravascular ultrasound has recently been introduced as a new technology in the assessment of coronary artery disease (2, 4, 5–14, 16, 17, 19, 20, 22). However, the experimental experience with intracoronary ultrasound for the diagnostic evaluation is limited (11, 12, 17). Furthermore, there is an obvious lack of experimental data in regard of correlation between the ultrasound image and the microscopic anatomy of the human coronary artery (2, 21). The results of transthoracic and transesophageal studies have demonstrated that echogenic contrast dye can improve image quality and provide higher accuracy of left and right ventricular measurements. Thus, the application of contrast dye might also improve the accuracy of lumen measurements of coronary arteries using the intravascular technique.

The current study was designed to address these questions in an in vitro-study in postmortem human coronary arteries comparing the ultrasound images with the histology of the corresponding segment. Aim of the study was to analyze the accuracy of intravascular ultrasound for determination of lumen area and diameter and to evaluate the impact of echogenic contrast dye on the accuracy of lumen measurements in coronary arteries.

Methods

Ultrasound transducer

We used a commercially available mechanical ultrasound imaging system, consisting of a rotating mirror reflecting the ultrasound beam of a 30 MHz transducer mounted on the tip of a 5F catheter (CVIS).

Material

Ultrasound was performed using postmortem coronary arteries of nine patients aged 60 ± 19 years (mean ± S.D., range: 38 to 86 years). Eight patients were male, one female. Arterial specimens were collected during autopsy, and were studied fresh, without fixation within 6 h after necropsy. The arteries were isolated sparing the adventitia and surrounding soft tissue to simulate in situ conditions. Side branches were tied off with sutures. Afterwards, the vessels were mounted on a special device (Fig. 1) and perfused by physiologic saline solution. Intraluminal pressure was maintained constant at 70 mm Hg. Eighty segments of 20 coronary arteries were visualized by two different mechanical ultrasound devices and compared to the respective histological findings.

Ultrasound imaging

The intravascular ultrasound catheter was inserted and aligned coaxially to the arterial segment. The total arrangement (Fig. 1) was then immersed in a physiologic saline bath which was kept at room temperature. To simulate in vivo conditions, images were not selected from sites in which image quality was highest, but were taken randomly at 0.5 cm intervals along the entire length of the artery. Gain, contrast, and other parameters of pre- and postprocessing were adjusted for each image to provide optimal image quality. The position of the ultrasound catheter was marked within the surrounding soft tissue by needles for each measuring site, thus allowing the exact circular and longitudinal orientation.

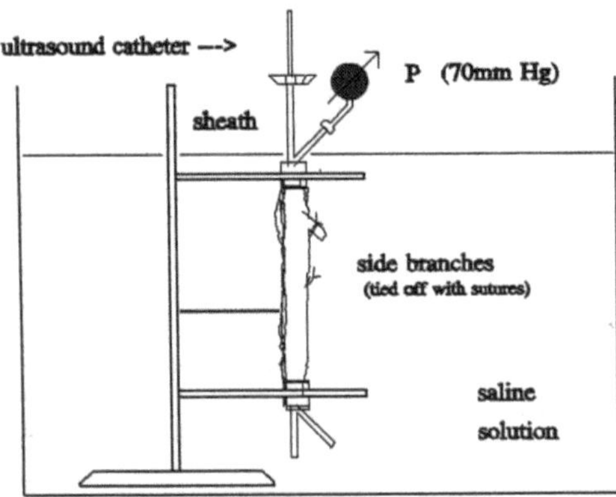

Fig. 1. Experimental setting.

Histology

After the ultrasound imaging procedure arteries were pressure (70 mm Hg) perfusion-fixed in 10 % neutral-buffered formalin and 24 h later placed in a standard decalcification solution. After embedding (paraffin), histological sections were taken at the sites earlier marked by needles. All sections were stained with hematoxilin-eosin stains.

Analysis of ultrasound images

The ultrasound images were analyzed by two experienced investigators using stop-frame images; in cases of disagreement, analysis was repeated. Lumen diameter was defined as the arithmetical mean of the largest and smallest diameter. Luminal area was determined by planimetry after definition of the luminal border of intima. Measurements were repeated after application of a bolus of 2 ml of echogenic contrast dye (SH U 508 = Laevovist, Schering AG, Berlin, FRG) during the washout phase. Perfusion pressure was maintained constant at 70 mm Hg during this procedure.

The diameter of the ringdown effect was 2.0 ± 0.04 mm (n = 20). In vessels with a lumen of less the ringdown diameter (n = 15) no reliable correlation could be found for area measurements even with contrast dye.

All measurements were performed online, using the software of the ultrasound device and were repeated offline for determination of intraobserver variability, using a Cardio-200 image analysis device (Kontron, Munich, FRG). Intraobserver variability was 0.04 ± 0.27 mm^2 for area measurements; no observer – dependant bias was found.

Definition of the vessel border by ultrasound along more than 270° of the circumference was possible in 78 segments.

Analysis of histology

The histologic specimens were analyzed under light microscopy by a cardiovascular pathologist, who was unaware of the ultrasound wall characterization and measurements. Morphometry was performed by a semiautomatic analysis device with previously validated software using a Reichert-Jung Mod. Univar microscope with tracing device and a modified digitizer (Summagraphics) with light point cursor (MTS GmbH, Tübingen, FRG) connected to a Commodore personal computer. Analysis was based on the same definitions as for the ultrasound imaging.

Statistical Analysis

Correlations between intravascular ultrasound and histology were determined by a linear regression analysis. The level of significance was tested, considering a probability value of less than 0.05 as significant. Additionally, subgroups were constituted according to the histological lumen diameter (smaller vessels: lumen diameter ≤ 2.5 mm, n = 36; larger vessels: lumen diameter > 2.5 mm, n = 44). Chi-square test and Student's paired *t*-test were performed using a statistical programm (PlotIT, TM) and Z-values were calculated to compare different correlations. For area measurements intraobsever variability was determined.

Results

Luminal diameter of the analyzed vessels was 2.6 mm ± 0.6 mm (mean ± S.D., range 1.7 to 4.4 mm) measured by histology and 2.8 mm ± 0.7 mm for ultrasound measurements. The correlation coefficient was 0.85 (Fig. 2). Thus, diameter calculated from ultrasound was 17 ± 21 % higher than the morphometric values (p = ns). According to the plot of the measurements of the perfusion-fixed arterial ring against the measurements of the corresponding histological section (Fig. 3), we could confirm the validity of our histological findings.

Luminal area determined by histology was 5.8 ± 3.5 mm² and 7.2 ± 3.5 mm² by ultrasound. Correlation coefficient between intravascular ultrasound and histology was r = 0.89 (Fig. 4). Again, a slight overestimation of luminal area was found, which was 37 ± 48 % (p < 0.05), leading to an overestimation of area in 65 segments (81 %). Subgroup analysis revealed that in vessels with a lumen diameter ≥2.5 mm a good correlation between ultrasound and histology could be found (r = 0.87). In contrast, for smaller sized vessels (lumen diameter <2.5 mm) correlation was significantly (p <0.02) reduced (r = 0.48).

Impact of contrast medium on intracoronary ultrasound: Luminal borders could be defined only after application of contrast dye in 2 %. In 30 cases (38 %) area measurements were not identical with and without of contrast dye. The number of area measurements which had to be corrected after application of contrast dye by more than 15 % was low (n = 10 [13 %]), but in all of these cases, the difference between ultrasound and histological measurements could be reduced (Table 1). Correlation coefficient for area measurements comparing ultrasound and histology could be increased from 0.85 to 0.89 (p = NS). In smaller size vessels, however, a relationship between ultrasound and histology was found only with contrast dye (without: p = NS; with contrast dye: r = 0.48, p <0.02).

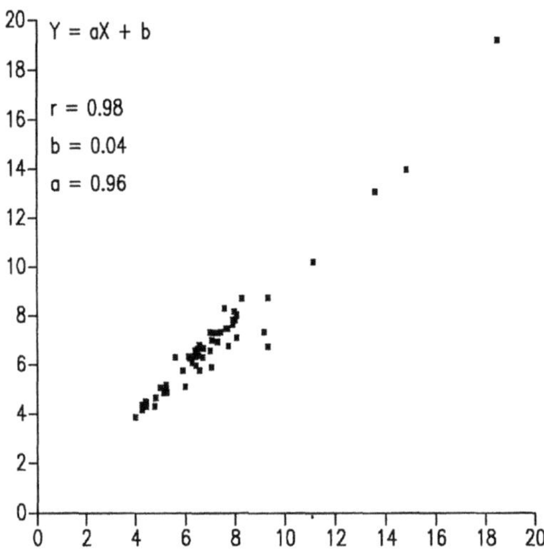

Fig. 2. Plot of the gross vessel diameter of the perfusion-fixed coronary arteries before and after histological procedures.

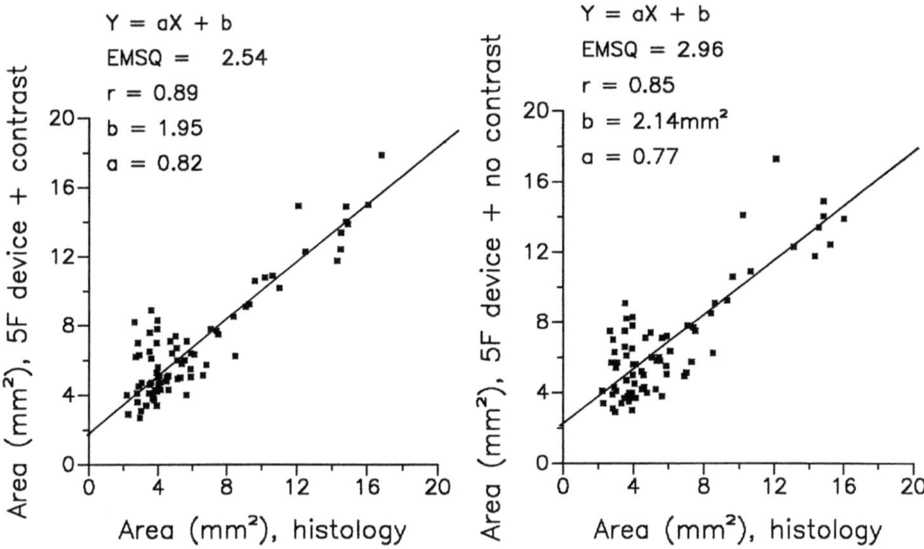

Fig. 3. Plot of lumen area measurements compared to histological measurements at the identical site a) with and b) without contrast dye.

Table 1. Impact of contrast dye on the accuracy of lumen measurements by intravascular ultrasound.

Number of lumen area measurements possible without contrast dye	Measurements not identical with and without contrast dye	Deviation of measurements with and without contrast dye > 15 %	From these: Contrast dye reduces deviation between ultrasound and histology	Measurements only possible after contrast dye*)
98 % (n = 78)	38 % (n = 30)	13 % (n = 10)	100 % (10 of 10)	3 % (n = 2)

*) All vessels of this group had a lumen diameter of ≤2.5 mm. Lumen and area measurements were possible in all 80 segments after application of contrast dye

Discussion

In concordance with various other authors (2, 5, 9, 13, 20) a slight overestimation of vessel lumen by intravascular ultrasound in comparison to histology was documented. This effect may be due to shrinkage by tissue processing, and thus, intracoronary ultrasound might much better reflect the real situation as an underestimation of lumen area by histology by 27 %. Siegel et al. (16) also reported a smaller

133

lumen area determined by histology comparing fresh tissue specimens to the findings after tissue processing (mean of difference: 26 %).

We found a good correlation between intravascular ultrasound and histology in quantification of *lumen area* (r = 0.88). These results are in agreement with the results reported by Potkin (4) and Tobis (1), both using an alternative 5F ultrasound system based on comparable techniques (r = 0.85 [4], r = 0.88 [1]).

In concordance to various other authors (2, 4, 9, 17), there was a very low intraobserver variability of <0.05 mm, underscoring that the results of intracoronary lumen assessment are highly reproducible.

In studies dealing with very large arterial segments (3) correlation coefficients (up to 0.97) are reported which are significantly higher than those obtained in intracoronary investigations. Especially in small diameter coronary arteries, no reliable relationship between ultrasound and histological measurements was found. These results may be due to a reduced resolution capability of the system in the near field and due to ringdown effects.

No systematical studies on the impact of contrast medium for exact definition of luminal borders have been reported. As demonstrated in this investigation, application of contrast medium allowed determination of lumen area in all cases and lead to an improvement of quantitative measurements in 38 % of all segments. Even in larger sized vessels an improvement of image quality could be achieved. Subgroup analysis, however, revealed that the effect of contrast dye was most important in small diameter vessels. The improved imaging quality in the near field may be of great promise; especially for the assessment of intraluminal structures further clinical studies should be performed.

In conclusion, there are considerable limitations in the accuracy of ultrasound measurements in the near field. Thus, further improvement of intracoronary ultrasound devices is mandatory. However, with the use of the currently available systems, additional application of contrast dye significantly improved the accuracy of luminal measurements, especially in smaller diameter vessels.

References

1. Baker JR (1958) The reactions of fixatives with tissue cells. Methods of research. In: Principles of Biological Microtechnique. London: Methuen 76–88
2. Borst C, Savalle LH, Smits PC, Post MJ, Gussenhoven WJ, Bom N (1991) Imaging of postmortem coronary arteries by 30 MHz intravascular ultrasound. International Journal of Cardiac Imaging 6: 239–246
3. Davidson CJ, Sheikh KH, Harrison JK (1990) Intravascular ultrasonography versus digital subtraction angiography: a human in-vivo comparison of vessel size and morphology. J Am Coll Cardiol 16: 633–636
4. Gussenhoven EJ, Essed CE, Lancee CT, Mastik F, Frietman P, Van Egmond FCm, Reiber J, Bosch H, Van Hurk H, Roelandt J, Bom N (1989) Arterial wall characteristics determined by intravascular imaging: An in vitro study. J AM Coll Cardiol 14: 947–942
5. Gussenhoven WJ, Essed CE, Frietmann P, Mastik F, Lancée C, Slager C, Serruys P, Gerritsen P, Pietermann H, Bom N (1989) Intravascular echocardiographic assessment of vessel wall characteristics: Correlation with histology. International Journal of Cardiac Imaging 4: 105–116
6. Isner JM, Rosenfield K, Losordo DW, Rose L, Langevin RE, Razvi S, Kosovsky BD (1991) Combination balloon-ultrasound imaging for percutaneous transluminal angioplasty. Validation of imaging, analysis of recoil and identification of plaque fracture. Circulation 84: 739–754

7. Mallery JA, Tobis JM, Griffith J, Gessert J, McRae M, Moussabeck O, Bessen M, Moriuchi M, Henry WL (1990) Assessment of normal and arteriosclerotic arterial wall thickness with an intravascular ultrasound imaging catheter. Am Heart J 6: 1392–1499
8. Neville JR, Yasuhara H, Watanabe BI, Canady J, Durán W, Hobson RW (1991) Endovascular management of arterial intimal defects: An experimental comparison by arteriography, angioscopy and intravascular ultrasonography. J Vasc Surg 13: 496–502
9. Nishimura RA, Edwards WD, Warnes CA, Reeder GS, Holmes DR, Tajik AJ, Yock PG (1990) Intravascular ultrasound imaging: In vitro validation and pathologic correlation. J Am Coll Cardiol 16: 145–154
10. Nissen SE, Grines CL, Gurley JC, Sublett K, Haynie D, Diaz C, Booth DC, DeMaria AN (1990) Application of a new phased-array ultrasound imaging catheter in the assessment of vascular dimensions: In vivo comparison to cineangiography. Circulation 81: 660–666
11. Nissen SE, Gurley JC (1991) Application of intravascular ultrasound for detection and quantification of coronary arteriosclerosis. International Journal of Cardiac Imaging 6: 165–177
12. Nissen SE, Gurley JC, Grines CL, Booth DC, McClure R, Berk M, Fischer C, DeMaria AN (1991) Intravascular ultrasound assessment of lumen size and wall morphology in normal subjects and patients with coronary artery disease. Circulation 84: 1087–1099
13. Potkin BN, Bartorelli AL, Gessert JM, Neville RF, Almagor Y, Roberts WC, Leon MB (1990) Coronary artery imaging with intravascular high frequency ultrasound. Circulation 81: 1575–1585
14. Rosenfield K, Losordo DW, Ramaswami K, Pastore JO, Langevin RE, Razvi S, Kosowsky BD, Isner JM (1991) Three-dimensional reconstruction of human coronary and peripheral erteries from images recorded during twodimensional intravascular ultrasound examination. Circulation 84: 1938–1956
15. Siegel RJ, Swan K, Edwalds G, Fishbein MC (1985) Limitations of postmortem assessment of human coronary artery size and luminal narrowing: Differential effects of tissue fixation and processing on vessels with different degrees of arteriosclerosis. J Am Coll Cardiol 5: 342–346
16. Siegel RJ, Ariani M, Fishbein MC, Chae JS, Park JC, Maurer G, Forrester JS (1991) Histopathologic validation of angioscopy and intravascular ultrasound. Circulation 84: 109–117
17. St. Goar F, Pinto FJ, Alderman EL, Fitzgerald PJ, Stadius ML, Popp RL (1991) Intravascular ultrasound imaging of angiographically normal coronary arteries: An in vivo comparison with quantitative angiography. J Am Coll Cardiol 18: 952–958
18. Stowell RE (1941) Effect on tissue volume of various methods of fixation, dehydration and embedding. Stain technology 16: 67–83
19. Tobis JM, Mallery JA, Gessert J, Griffith J, Mahon D, Bessen M, Moriuchi M, McLeay L, McRae M, Henry WL (1989) Intravascular ultrasound cross-sectional arterial imaging before and after balloon angioplasty in vitro. Circulation 80: 873–882
20. Tobis JM, Mallery JA, Mahon D, Lehmann K, Zalesky P, Griffith J, Gessert J, Moriuchi M, McRae M, Dwyer ML, Greep N, Henry WL (1991) Intravascular ultrasound imaging of human coronary arteries in vivo – Analysis of tissue charactization with comparison to in vitro histological specimens. Circulation 83: 913–926
21. Waller BF (1989) Anatomy, histology and pathology of the major epicardial coronary arteries relevant to echocardiographic techniques. J Am Soc Echo 2: 232–252
22. Yock PG, Fitzgerald PJ, Sudhir K, Linker DT, White W, Ports A (1991) Intravascular ultrasound imaging for guidance of atherectomy and other plaque removal techniques. International Journal of Cardiac Imaging 6: 179–189

Author's address:
Karl M. Schmid, MD
Department of Cardiology
University of Tübingen
Otfried-Müller-Str. 10
D-72076 Tübingen
FRG

Therapeutic approaches to the control of coronary atherosclerosis

W. G. Nayler

Department of Medicine, University of Melbourne, Austin Hospital, Heidelberg, VIC, Australia

Summary: Atherosclerosis is a multifactorial disease which culminates in the ruptured plaque seen at autopsy. Hypercholesterolaemia, subintimal accumulation of lipid, monocyte adhesion followed by penetration across the endothelium, the conversion of monocytes to macrophages and smooth muscle cell proliferation and migration are some of the events involved in the early stages of lesion formation. Late events include the formation of excess ground substance and collagen, and the formation of the fibrotic cap. Young lesions tend to be more fragile than "old" calcified lesions, and it is these young lesions which rupture, haemorrhage and provide anchor points for platelets.

Therapeutic interventions aimed at controlling lesion formation include those which reduce risk factors, including hypertension as well as those which interfere with the cascade of events involved in lesion formation. Agents which lower plasma cholesterol provide one approach. Another approach is to use calcium antagonists which not only lower blood pressure, but also directly interfere with some of the metabolic events involved in lesion formation.

Key words: Calcium antagonists – nifedipine – nisoldipine – cholesterol – atherogenesis – oxyradicals

Introduction

During the past decade considerable advances have been made in our understanding of the events which culminate in the development of an atherosclerotic lesion. Recognized risk factors – including hypertension, diabetes, smoking and abnormally high levels of plasma cholesterol – have been documented, as has the sequence of physical and biochemical events which, according to our current concepts of the disease process, are involved in the formation of the lesion. Events which are thought to be involved include hypercholesterolaemia, lipid accumulation in the subintimal space and their subsequent oxidation, monocyte adhesion to and penetration across the endothelium, smooth muscle cell proliferation and migration and excess formation of collagen, matrix material and ground substance (3, 9, 18).

Since the fissuring (or rupturing) of an atherosclerotic lesion in the coronary vasculature is a major cause of myocardial ischaemia, and even death, it is becoming increasingly important to identify the cellular events which contribute to lesion formation (18), as well as the events which cause the lesion to rupture and haemorrhage (3). The most recent data indicates that it is the young lesions which fracture and haemorrhage, a property which has allowed some investigators to argue that it is the developing lesions which should be targeted for treatment.

Therapeutic regimes developed for the management of patients "at risk" can be divided into two categories – those regimes which aim to reduce the incidence or severity of risk factors, and those which aim at interrupting the cascade of events involved in the formation of the lesions. This cascade includes a raised plasma cholesterol, and the penetration of lipid across the endothelium into the subintimal space where it becomes oxidized. This is followed by the adhesion of monocytes to the extracellular surface of the endothelium and the penetration of these cells into the subintimal space where they convert to macrophages which proceed to accumulate the now oxidized lipid (low-density lipid), thereby becoming foam cells. Platelets also adhere to the endothelial surface, possibly because in the presence of a raised plasma level of low density lipids (LDL) this surface becomes "sticky". These adherent platelets, together with the resident macrophages now release "growth factors" which act on the smooth muscle cells, causing them to proliferate and migrate into the intima. By this time the lesion is beginning to increase in bulk, pushing its way into the endothelium until ultimately it, together with the now ruptured endothelial layer, begins to calcify and increase in bulk (9). It is at this stage that an "adult" plaque can be identified at angiography. It is also at this stage that the "plaque" is likely to rupture – thereby presenting the platelets with a surface area on which they can aggregate (3, 16, 18).

Many of the events which are involved in the "atherogenic cascade" are known to be Ca^{2+}-dependent. Such events include the release of growth factors (15), the migration of smooth muscle cells (16), and the stimulation of collagen synthesis (6). The Ca^{2+}-dependent nature of these events, together with the presence of calcified material in the adult ruptured plaques (2) has provided the basis for investigations aimed at proving whether calcium antagonists are effective antiatherogenic agents.

The following experiments were undertaken to demonstrate the effectiveness of nifedipine and nisoldipine in this regard.

Methods

Nifedipine Series

Young male New Zealand white rabbits were used for these experiments. The rabbits were housed in a temperature – and humidity – controlled animal house, and they weighed between 1.5 and 2 kg at the start of the experiments. The rabbits were stabilized for 7 days before starting the experimental protocol, during which time they were fed standard rabbit pellets and water ad libitum. Foods intake was measured on a daily basis. After this 7 days of stabilization the rabbits were assigned at random to one of the following regimes:

I.   standard pellets plus placebo;

II. standard pellets plus nifedipine (0.5–5 mg/kg/day);

III. cholesterol-enriched pellets (5 % cholesterol plus 2.5 % peanut oil, by weight) plus placebo; and

IV. cholesterol-enriched pellets as above plus nifedipine (0.5–5 mg/kg/day).

The cholesterol-enriched diet was prepared in the laboratory, by dissolving the cholesterol in peanut oil, and ether, and adding the resultant mixture to the pulverized pellets. The resultant homogenate was air-dried overnight under fluorescent lights. The nifedipine, after being suspended in ethanol, was added to the pellets which were again dried, but under sodium lighting, to avoid photolysis of the drug.

Treatment was for 12 weeks, during which time the food intake was adjusted on a daily basis to ensure that the required amount of nifedipine was consumed. After 12 weeks of treatment the rabbits were killed by cervical dislocation, after blood had been taken from the central ear artery for cholesterol determination. The thoracic and abdominal aorta were removed, and either stained with Sudan IV for sudanophilic lesions, or assayed, for cholesterol (21) or calcium content. Calcium content was measured by atomic absorption spectrometery as previously described (1), using HNO_3, digests of abdominal or thoracic aorta dried to constant weight.

The percentage of aortic surface occupied by sudanophilic lesions was determined by computer-assisted analysis of variance, with the Bonferroni adjustment for multiple comparisons; $p = 0.05$ was taken as the limit of significance. Results are presented as mean $\pm$ SEM of n experiments.

Nisoldipine series

In a parallel series of experiments nisoldipine (0.5–5.0 mg/kg/day) was used instead of nifedipine.

Results

Effect of nifedipine or nisoldipine on plasma cholesterol levels after
12 weeks of treatment

Plasma taken from placebo-fed rabbits contained 2.2 ± 0.1 mmol cholesterol per litre (mean $\pm$ SEM) of 12 experiments). By contrast rabbits maintained on the cholesterol rich diet for 12 weeks contained 108 ± 6. mmol/litre cholesterol (mean $\pm$ SEM, 12 experiments). Adding 0.5–5.0 mg/kg/day of either nifedipine or nisoldipine to the diet of the cholesterol-fed rabbits failed to cause any significant change in the plasma cholesterol levels (82 ± 7.6 mmol/litre for the nifedipine series and 89 ± 8.2 mmol/litre for the nisoldipine group – where each result is the mean $\pm$ SEM of 12 experiments over a treatment period of 12 weeks).

Effect of nifedipine or nisoldipine on tissue Ca^{2+} and cholesterol content of aorta of
cholesterol-fed rabbits

The effect of 12 weeks treatment with either nifedipine (0.5–5.0 mg/kg/day) or nisoldipine (0.5–5.0 mg/kg/day) on the cholesterol and Ca^{2+} content of the abdominal and thoracic aorta is shown in Table 1. For both calcium antagonists and in both the abdominal and thoracic aortae there is a dose-dependent reduction in aortic cholesterol and Ca^{2+}, the reductions being significant even at the lowest dose level used here – 0.5 mg/kg/day.

Table 1. Effect of 12 weeks treatment with either nifedipine or nisoldipine on abdominal and thoracic aortic Ca^{2+} and cholesterol in cholesterol-fed rabbits.

	Thoracic aorta	Abdominal aorta
Series A: Nifedipine (mg/kg/day)	Cholesterol (mg.g dry weight^{-1})	
0	78 ± 9	112 ± 11
0.5	56 ± 7 (p < 0.05)	79 ± 8 (p < 0.01)
1.0	49 ± 12 (p < 0.01)	62 ± 10 (p < 0.01)
5.0	41 ± 6 (p < 0.01)	51 ± 11 (p < 0.01)
	Ca^{2+} (μg.g dry weight^{-1})	
0	126 ± 14	158 ± 25
0.5	71 ± 12 (p < 0.01)	92 ± 21 (p < 0.01)
1.0	63 ± 14 (p < 0.01)	71 ± 11 (p < 0.01)
5.0	41 ± 11 (p < 0.01)	56 ± 12 (p < 0.001)
Series B: Nisoldipine (mg/kg/day)	Cholesterol (mg.g dry weight^{-1})	
0	83 ± 12	92 ± 9
0.5	41 ± 6 (p < 0.01)	62 ± 11 (p < 0.01)
1.0	37 ± 7 (p < 0.01)	41 ± 9 (p < 0.01)
5.0	21 ± 9 (p < 0.001)	32 ± 8 (p < 0.001)
	Ca^{2+} (μg.g dry weight^{-1})	
0	118 ± 11	126 ± 14
0.5	60 ± 5 (p < 0.01)	72 ± 11 (p < 0.01)
1.0	49 ± 12 (p < 0.001)	61 ± 19 (p < 0.01)
5.0	42 ± 6 (p < 0.001)	53 ± 11 (p < 0.001)

Each result is the mean ± SEM of eight separate experiments. Treatment period was for 12 weeks. Estimations of tissue Ca^{2+} and cholesterol content were performed as described in the text in duplicate. Tests of significance relate to the significance of the difference relative to the values obtained in the absence of either nifedipine or nisoldipine at the indicated doses.

Effect of nifedipine or nisoldipine on the formation of sudanophilic positive (atherogenic) lesions in the abdominal and thoracic aortae of cholesterol-fed rabbits

The effect of 12 weeks treatment with either nifedipine (0.5–5.0 mg/kg/day) or nisoldipine (0.5–5.0 mg/kg/day) on the incidence of sudanophilic lesions in the abdominal and thoracic aortae of these cholesterol-fed rabbits is shown in Table 2. Both calcium antagonists produced a dose-dependent reduction in the incidence of lesion formation, over a dose range which (see above) had no effect on the plasma cholesterol levels.

Discussion

These results confirm the ability of nifedipine (5, 10) to slow the progression of atherosclerosis in cholesterol-fed rabbits, and establish that another dihy-

Table 2. Effect of 12 weeks treatment with either nifedipine or nisoldipine on the incidence of sudanophilic positive lesions in the abdominal and thoracic aortae of cholesterol-fed rabbits.

| | % Incidence of sudanophilic lesions | |
	Thoracic Aorta	Abdominal Aorta
Nifedipine series (mg/kg/day)		
0	48 ± 9	70 ± 13.5
0.5	30.5 ± 8 (p < 0.05)	41 ± 8 (p < 0.01)
1.0	24.5 ± 10 (p < 0.05)	32 ± 10 (p < 0.01)
5.0	12.6 ± 9 (p < 0.01)	12 ± 4 (p < 0.01)
Nisoldipine series (mg/kg/day)		
0	58 ± 12	73 ± 8
0.5	31 ± 8 (p < 0.01)	52 ± 12 (p < 0.05)
1.0	22 ± 5 (p < 0.01)	41 ± 11 (p < 0.01)
5.0	11 ± 8 (p < 0.01)	28 ± 6 (p < 0.001)

Each result is mean SEM of eight experiments. Treatment period was for 12 weeks. % sudanophilic lesions refers to the percentage of the surface area of the vessel wall occupied by sudanophilic positive lesions. Tests of significance relate to results obtained in the absence of nisoldipine.

dropyridine-based antagonist (nisoldipine) mimics nifedipine in this regard. Since nifedipine has been shown to exert a similar effect in clinical trials on patients at risk from atherogenesis (7) it might be assumed that nisoldipine will also be effective in man – but clinical trials are needed before such a conclusion can be unequivocally accepted.

Whilst these results establish the efficacy of dihydropyridine-based calcium antagonists in so far as they slow the growth of atherogenic lesions in cholesterol-fed rabbits, some investigators have been cautious in applying the results to naturally-occurring atherosclerosis in man. Such caution is reasonable because whereas in man the lesion develops slowly and in the presence of other risk factors in addition to hypercholesterolaemia, in the rabbit model the lesion develops rapidly, and the plasma cholesterol levels are greatly increased. Nevertheless, other investigators have provided evidence which points to why those drugs are effective. For example, drugs of this type have a direct inhibitory effect on smooth muscle cell proliferation (11, 12). They also inhibit neutrophil and macrophage chemotaxis (11), and inhibit the hydrolysis of lipoproteins (14) and cholesterol ester metabolism (13, 19, 20). In addition, they inhibit matrix synthesis and platelet aggregation (22), as well as reducing one of the major risk factors – that of hypertension. Clearly, however, they do not lower plasma lipids.

Certainly there are other approaches to the management of patients with atherogenesis. For example, pharmacological interventions which are aimed at lowering serum cholesterol can be used. Such agents include the HMGCoA reductase inhibitors. These act by decreasing the intrahepatic concentrations of cholesterol, to stimulate LDL receptor activity. Another approach is to hasten the clearance of triglycerides from the circulation – by using agents which stimulate lipoprotein lipase activity. Alternatively, agents which inhibit the oxidation of the low density lipoproteins, or which prevent macrophages, platelets and endothelial cells from becoming "sticky" might be developed for use in this condition. For the purposes of this present paper, however, it is sufficient to conclude that the dihydropyridine-based calcium antagonists do provide an effective form of therapy which, in the case of nifedipine (4, 7, 8) and nicardipine (22) has been shown to be effective in man.

Acknowledgements. These investigations were carried out during the tenure of a grant from the National Health and Medical Research Council of Australia.

References

1. Daly MJ, Elz JS, Nayler WG (1987) Contracture and the calcium paradox in the rat heart. Circ Res 61: 560–569
2. Fleckenstein A, Frey M, Thimm F, Fleckenstein-Grun G (1990) Excessive mural calcium overload a predominant causal factor in the development of stenosing coronary plaques in humans. Cardiovasc Drugs Ther 4: 1005–1014
3. Fuster V, Stein B, Ambrose JA, Badimon I, Badimon JJ, Chesbro JH (1990) Atherosclerotic plaque rupture and thrombis: evolving concepts. Circulation 82 (Suppl II): II-47–II-59
4. Gottlieb SO, Brinker JA, Mellitis ED, Aschuff SC, Baughman KL, Traill TA, Weiss JL, Reitz BA, Weisfeldt M, Gerstenblith G (1989) Effect of nifedipine on the development of coronary bypass graft stenosis in high-risk patients: a randomized double-blind, placebo-controlled trial. Circulation 80 (Suppl 2) II-228 (abstract)
5. Henry PD, Bentley KI (1981) Suppression of atherogenesis in cholesterol-fed rabbits treated with nifedipine. J Clin Invest 68: 1366–1369
6. Kramsch DM, Aspen AJ, Rozier LJ (1981) Atherosclerosis: prevention by agents not affecting abnormal levels of blood lipids. Science 213: 1511–1512
7. Lichtlen PR, Hugenholtz PG, Rafflenbeul W, Hecker H, Jost S, Deckers JW (1990) Retardation of angiographic progression of coronary artery disease by nifedipine. Lancet 335: 1109–1113
8. Loaldi A, Polese A, Montorsi P, De Cesare N, Fabbiocchi F, Ravagnani P, Guazzi M (1989) Comparison of nifedipine, propranolol and isosorbide dinitrate on angiographic progression and regression of coronary artery narrowing in angina pectoris. Am J Cardiol 64: 433–439
9. Nayler WG (1991) Molecular mechanisms involved in the anti-atherogenic effect of the calcium antagonists. In: Nayler WG, The Second Generation of Calcium Antagonists, Springer Verlag, Berlin, pp 139–151
10. Nayler WG, Pangiotopoulos S (1986) Calcium antagonism. In: Kelly DT (ed) Proceedings of II Asian Pacific Adalat Symposium, ADIS Press, New Zealand, pp 3–11
11. Neuser D, Rosen B (1990) Calcium antagonists and the proliferation of vascular smooth muscle cells. Satellite Symposium of 13th ISH meeting Montreal "Antiatherosclerotic Potential of Calcium Antagonists" pp 9–10
12. Nilsson I, Sjolund M, Palmberg L, Von Euler AM, Jonzon B, Thyborg I (1985) The calcium antagonist nifedipine inhibits arterial smooth muscle cell proliferation. Atherosclerosis 58: 109–112
13. Orekov AN, Tertov VU, Khashimov KA, Kudryashou SA, Smirnov VN (1986) Anti-atherosclerotic effects of verapamil in primary culture of human aortic intimal cells. J Hypertens 4 (Suppl 6): S 153–155

14. Ranganthan S, Jackson RL (1984) Effect of calcium channel blocking drugs on lysosomal function in human skin fibroblasts. Biochem Pharmacol 33: 2377–2382
15. Ross R (1981) Atherosclerosis: a problem of the biology of arterial wall cells and their interactions with blood components. Arteriosclerosis 1: 293–311
16. Ross R (1986) The pathogenesis of atherosclerosis – an update. New England J Med 314: 488–500
17. Ross R, Glomset JA (1976) The pathogenesis of atherosclerosis. New England J Med 295: 369–377
18. Schwartz CJ, Kelley J, Nerem RM, Sprague EA (1989) Pathophysiology of the atherogenic process. Am J Cardiol 64: 236–305
19. Stein O, Stein Y (1987) Effect of verapamil on cholesteryl ester hydrolysis and re-esterification in macrophages. Atherosclerosis 7: 578–584
20. Stein O, Halpenis G, Stein Y (1987) Long term effects of verapamil on aortic smooth muscle cells cultured in the presence of hypercholesterolaemic serum. Atherosclerosis 7: 585–592
21. Tonks BD (1967) The estimation of cholesterol in serum: a classification and critical review of methods. Clin Biochem 1: 12–29
22. Ware JA, Johnson PC, Smith M, Salzman EW (1986) Inhibition of human platelet aggregation and cytoplasmic calcium response by calcium antagonists. Circ Res 59: 39–42
23. Waters D, Lesperance J, Francetich M, Causey D, Theroux P, Chiang Y-K, Hudon G, Lemarbre L, Reitman M, Joyal M, Gosselin G, Dyrda I, Macer J, Havel RJ (1990) A controlled clinical trial to assess the effect of a calcium channel blocker on the progression of coronary atherosclerosis. Circulation 82: 1940–1953

Author's address:
Prof. W. G. Nayler, Ph. D.
Department of Medicine
The University of Melbourne
Austin Hospital
Heidelberg VIC 3084
Australia

Role of calcium in arteriosclerosis – Experimental evaluation of antiarteriosclerotic potencies of Ca antagonists

G. Fleckenstein-Grün*), F. Thimm, M. Frey, A. Czirfusz

Study Group for Calcium Antagonism

Summary: Chemical microanalyses of conventional human coronary artery plaques (stages I–III [WHO]) revealed the correlation between progressive mural Ca overload up to excessive degrees, and the severity of plaque formation, whereas only small amounts of cholesterol were found, even in complicated lesions. The pathogenetic role of Ca was tested in three types of experimental arteriosclerosis and atheromatosis, using Ca antagonists (verapamil, nitrendipine, diltiazem) as research tools: 1) The Ca type, in vitamin D_3 plus nicotine-treated rats; 2) the cholesterol type, in cholesterol-fed New Zealand rabbits; 3) mixed types, in SHRs and NaCl-fed Dahl-S rats. Types (1) and (3) were demonstrated to be governed by a progressive arterial Ca uptake that could be established already in early lesions. The increased mural Ca supply promoted cellular necroses, migration, and proliferation, as well as calcification and degradation of elastic fibers. Ca antagonists prevented the increased Ca incorporation into arterial walls and inhibited the development of experimental arterioscleroses of types (1) and (3). Ca antagonists did not protect coronary arteries of cholesterol-fed rabbits (type [2]) from occlusive cholesterol accumulation. The data suggest an important pathogenetic role of Ca and pronounced antiarteriosclerotic potencies of Ca antagonists in Ca-dominated types of experimental arteriosclerosis. The significance of the present results for pathophysiology and therapy of conventional human arteriosclerosis remains to be clarified.

Key words: Calcium antagonists – vasoprotection – conventional human arteriosclerosis – cholesterol hypothesis of arteriosclerosis

Introduction

Human arteriosclerosis develops slowly, over decades, as "response to injury" (25). Injury of the arterial wall is said to occur from endothelial damage. In response, various cell types, i.e., endothelial cells, platelets, vascular smooth muscle cells (VSMCs), monocytes/macrophages, and T-lymphocytes are discussed to interact via a multitude of cytokines, growth factors, endothelins, matrix molecules, etc. Hereditary predisposition, diabetes and hypertension, as well as toxic effects of free radicals, oxidatively modified lipoproteins, and nicotine obviously promote these complex interactions (20).

The cholesterol hypothesis of arteriosclerosis has dominated the scientific literature (2), whereas the wellknown occurrence of arterial calcium (Ca) overload was not taken for more than a trivial phenomenon without any pathogenetic importance.

---

*) Dedicated to my husband Prof. Dr. Dr. med. h.c. mult. Albrecht Fleckenstein (1917–1992)

The role of Ca in arteriosclerosis appeared in a new light, when we demonstrated that specific Ca antagonists prevent Ca overload in various models of experimental arteriosclerosis and protect arterial walls from Ca-mediated arteriosclerotic alterations (7, 8, 10).

The aim of this report is to present a quantitative microanalysis of Ca and lipid accumulation in different stages of conventional human coronary artery plaques. Moreover, the pathogenetic importance of Ca will be discussed, using specific Ca antagonists (verapamil, nitrendipine, diltiazem) as research tools in three different types of experimental arteriosclerosis or atheromatosis, respectively.

Methods

Human coronary artery specimens, consisting of intima plus media, were excised from autopsies, aged 40 to 90 years, in the Freiburg University's Institutes for Pathology and Forensic Medicine. Gross classification into the WHO stages I–III took place with the assistance of skilled pathologists. Stage I: Yellowish flat "Fatty streaks"; stage II: Whitish protrudent "fibrous plaques"; stage III: Severe, obstructive, bulky plaques, complicated by calcification, ulceration, necrosis and thrombosis (9).

Sprague-Dawley rats (200 g) were treated with either nicotine (2×25 mg/kg b.w./die p.o.), or vitamin D_3-hydrosol (one single i.m. injection of 300000 I.U./kg b.w.) for an observation period of 4 to 28 days; or with a combination of both vitamin D_3 and nicotine for 6 days. Diltiazem was injected s.c. (2×50 mg/kg b.w./die) simultaneously with the combined administration of vitamin D_3 and nicotine (11).

Sixteen-week-old male New Zealand rabbits (3–3.5 kg) were fed a cholesterol-rich diet (control diet for rabbits (C 2000, Altromin, Lage, FRG) containing 2 % cholesterol) until the age of 30–33 weeks. Simultaneously, pilules of Ca antagonists were administered orally twice daily (verapamil: 50–100 mg/kg b.w./die; nitrendipine: 12–15 mg/kg b.w./die).

Four-week-old spontaneously-hypertensive rats were treated with nitrendipine (300 mg/kg b.w./die p.o.) until the age of 82 weeks.

Six-week-old Dahl rats were fed a salt-rich diet (control diet for rats (C 1000, Altromin, Lage, FRG) containing 8 % NaCl) until the age of 16–24 weeks. Lovastatin (30–120 mg/kg b.w. p.o./die) and nitrendipine (10 mg/kg b.w. p.o./die) were given simultaneously in the diet.

Calcium and cholesterol contents of arterial walls were analyzed by atomic absorption spectroscopy (Perkin Elmer 3030 Atomic Absorption Spectrophotometer) or by gas chromatography (Varian 3400 Gas-Chromatograph) (9).

Serum lipids were analyzed by sequential ultracentrifugation technique.

Ca deposits were visualized with the von Kossa-staining technique in light, and with the K-pyroantimonate technique in electron microscopic studies.

In vivo net incorporations of ^{45}Ca and ^{14}C-cholesterol were measured in distal branches of superior mesenteric arteries of 20-month-old SHRs and Wistar rats by liquid scintillation (Canberra Packard, 1600 CA TRI-CARB Liquid Scintillation Analyzer), 6 h after i.p. injection of 30 μCi ^{45}Ca/kg b.w. and 24 h after i.p. injection of 20 μCi ^{14}C-cholesterol/kg b.w. and expressed as % of plasma activity ($= 100$ %).

Monolayers of rat's aortic media cells were cultured by standard methods in Ham's F-12 medium. Net ^{45}Ca incorporation was measured after addition of 2 μCi ^{45}Ca/l by liquid scintillation and calculated as ng Ca/mg protein (3).

Data were presented as mean ± S.E.M. Statistical analyses were carried out by Student's *t*-test.

Results

Progressive mural Ca accumulation – a primordial alteration already in early stages of conventional human coronary artery plaques

In conventional human coronary artery plaques, there was a positive correlation between the degree of Ca accumulation and the severity of arteriosclerosis (Fig. 1). In healthy coronary segments, the content (g/kg) of free and total cholesterol always surpassed that of Ca. Already in stage I plaques (so-called "fatty streaks"), the Ca content was 13 times, in stage II ("fibrous plaques") 24 times, and in stage III ("complicated lesions") 80 times the content of healthy coronary segments of the same age group. Since Ca was mainly present in the form of hydroxyapatite, the calculated proportion of Ca salts was approximately 40 % of the total plaque weight. Simultaneously, cholesterol was only increased by a factor of 3 in early lesions and doubled in fully developed plaques.

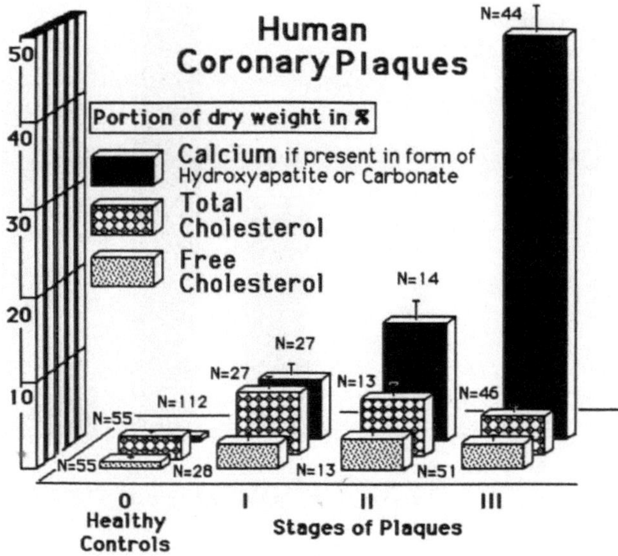

Fig. 1. Progressive rise in the proportion of Ca salts in human coronary artery plaques of increasing severity.

The role of Ca in three different types of experimental arteriosclerosis or atheromatosis respectively – antiarteriosclerotic efficacy of Ca antagonists

Calcium type of experimental arteriosclerosis

The simplest experimental way of producing tremendous degrees of arterial Ca overload, similar to Mönckeberg's arteriosclerosis, was by administration of an overdose of vitamin D_3 to rats or rabbits. The extracellular Ca concentration was elevated from 2.0 to 3.0 mmoles/l. In coronary arteries of rats, this arterial Ca overload was enormously potentiated by simultaneous application of nicotine. The combined treatment produced such deleterious degrees of coronary calcinosis that a 6-day period could not be survived. Prophylactic administration of diltiazem significantly suppressed arterial Ca overload and the animals survived (Fig. 2). In monolayers of rats' media cells, nicotine significantly stimulated the net Ca uptake. This nicotine effect was potentiated by an increase of the transmembrane Ca gradient. Conversely, specific Ca antagonists exerted protection from nicotine-induced intracellular Ca overload (Fig. 3). The electronmicrograph in Fig. 4 reflects the early lesions in a rat's coronary artery after injection of vitamin D_3. Ca was accumulated in media cells at the inner membrane layers and in intracellular organelles, particularly the mitochondria. Moreover, fragmentation of a heavily calcified elastic fibre is documented. Sometimes, but not always, fibrous intimal plaques developed in response to media injury.

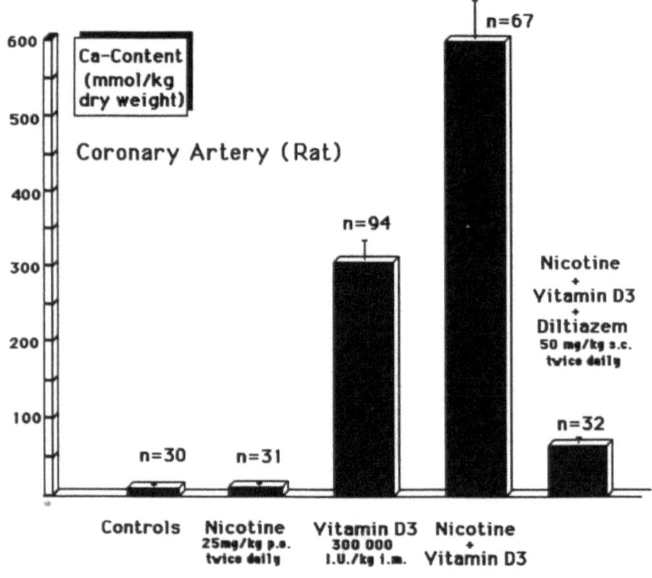

Fig. 2. Potentiation of vitamin D_3-induced arterial Ca overload by simultaneous oral administration of nicotine. Significant reduction of Ca overload by diltiazem (descending branches of the left coronary arteries of Sprague-Dawley rats).

148

Finally, in vitamin D_3 plus nicotine-treated rats, severe Ca overload completely destroyed arterial walls. Prophylactic treatment of rats with suitable Ca antagonists – as demonstrated here for diltiazem – prevented arterial Ca overload and maintained structural integrity (Fig. 5).

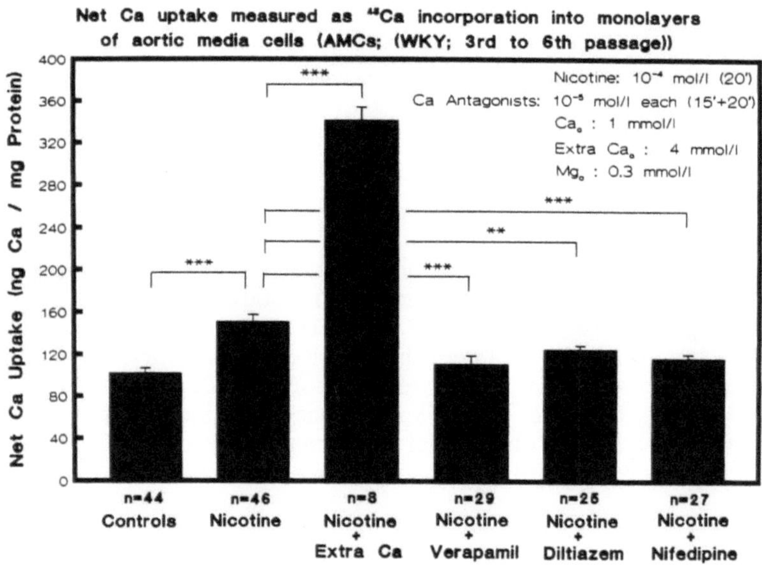

Fig. 3. Nicotine-induced increase of net Ca uptake into monolayers of rats' arterial media cells is potentiated by Ca and inhibited by Ca antagonists.

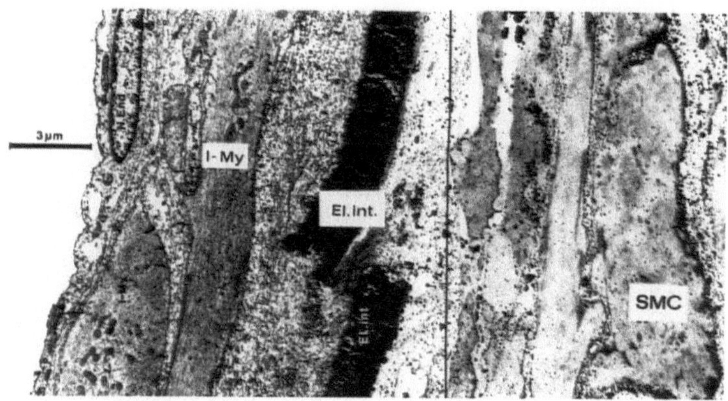

Fig. 4. Electron micrographs of early lesions in coronary arteries of vitamin D_3-treated rats (22). I-My: Myointimal cell. El. int.: Internal elastic membrane. SMC: Media Smooth muscle cell. Ca is visualized as black deposits with the potassium pyroantimonate technique (1:11200).

149

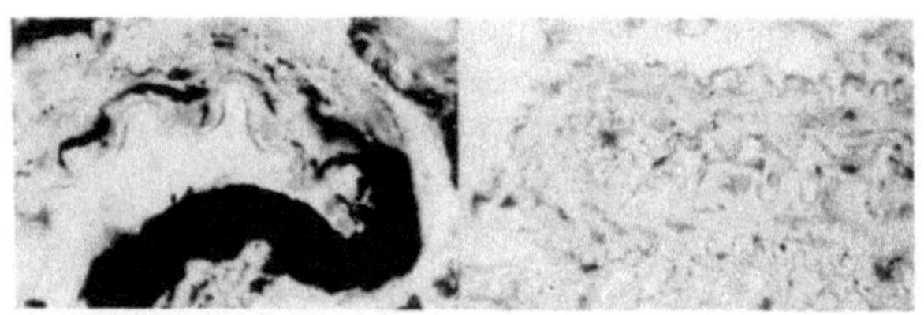

Fig. 5. Cross-sections of superior mesenteric arteries of rats (von Kossa's staining technique [1:250]). Ca is visualized as black deposits. Left: Massive arterial Ca overload after combined treatment with vitamin D_3 + nicotine. Right: Anticalcinotic vasoprotection with diltiazem (see Fig. 3).

Cholesterol type of experimental atheromatosis

Male New Zealand rabbits, fed a 2 % cholesterol diet, developed a generalized lipid storage disease within 14–17 weeks, if their serum total cholesterol was increased by a factor of roughly 40, i.e. from 52.4 ± 7.9 (n = 11) to 2171.9 ± 170.9 (n = 16) mg/dl. Early lesions in coronary arteries were characterized by a significant intimal accumulation of free and total cholesterol, whereas the coronary Ca content remained in the normal range (Fig. 6). Later on, following massive intima xanthomatosis, focal degenerative processes were observed at the intima-media border. They were characterized by calcified cell necrosis and calcification of elastic fibers. We treated choles-

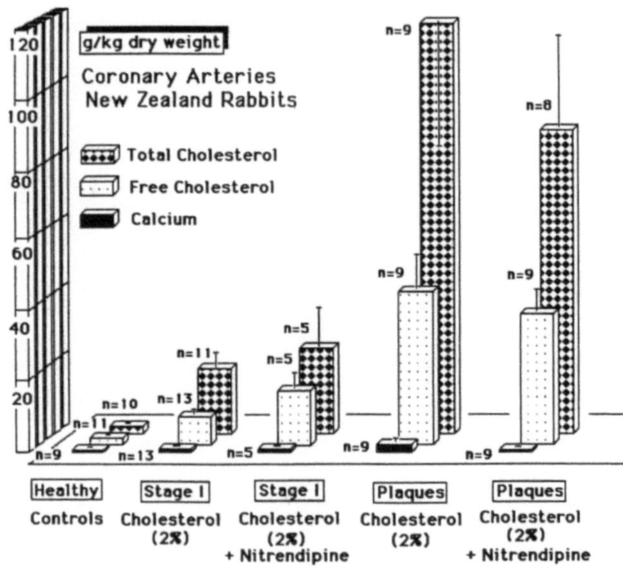

Fig. 6. Contents of Ca and free and total cholesterol of coronary arteries of male New Zealand rabbits. Stage I: Early lesions under a 2 % cholesterol diet (no visible alteration in stereomicroscopic inspection). Plaques: Occlusive cholesterol accumulations.

150

terol-fed rabbits with verapamil and nitrendipine. In very briefly summarizing our results: Ca antagonists did *not* significantly reduce cholesterol accumulation in aortae, coronary arteries, liver, kidney, and myocardium. In coronary arteries (Fig. 6), Ca antagonists, as for instance nitrendipine, did not prevent intimal foam cell accumulation or cholesterol incorporation, either in the early stage or in fully developed atheromata. But they significantly inhibited the secondary Ca overload and protected from Ca-mediated degeneration. However, Ca antagonists did not prevent multiple myocardial infarctions, caused in cholesterol-fed New Zealand rabbits by occlusive coronary cholesterol accumulation, whether the animals were treated with Ca antagonists or not (Fig. 7).

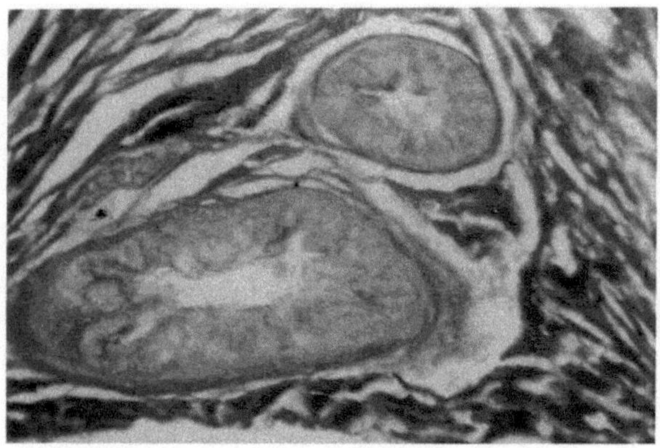

Fig. 7. Fully developed coronary cholesterol atheroma in a 16-week-old, 2 % cholesterol-fed New Zealand rabbit under long-term treatment with verapamil.

Mixed types of experimental arteriosclerosis

1. Experimental arteriosclerosis in spontaneously-hypertensive rats (SHR)

Arteriosclerotic lesions of SHR typically developed in distal branches of the superior mesenteric arteries. We classified them according to their severity (Fig. 8).

Stage I: Meandrous windings and irregular discrete increase in wall diameter. Stage II: Blue nodes with significant wall thickening. Stage III: White and yellow nodes with enormous increase in arterial diameter, tortuosity and wall thickening by a factor of more than 50. We analyzed the in vivo incorporation of ^{45}Ca und ^{14}C-cholesterol into these different stages (Fig. 9). Already in early lesions (Stage I) net ^{45}Ca incorporation (% of plasma activity [= 100 %]) was doubled (153.2 ± 39.8 [n = 10]) compared to normal (78.1 ± 1.9 [n = 10]), in type II plaques, Ca uptake was increased four-fold (322.6 ± 71.3 [n = 10]) and in type III 10-fold (900.8 ± 50 [n = 10]) above controls. In contrast, net ^{14}C-cholesterol only amounted 1.3 the most (from 66.1 ± 1.7 [n = 8] in healthy segments to 83.9 ± 11.6 [n = 8] in stages I and II). Ca antagonists, as for instance nitrendipine, from the beginning, prevented ^{45}Ca incorporation (115.5 ± 8.7 [n = 7]) and protected arteries from Ca-modulated

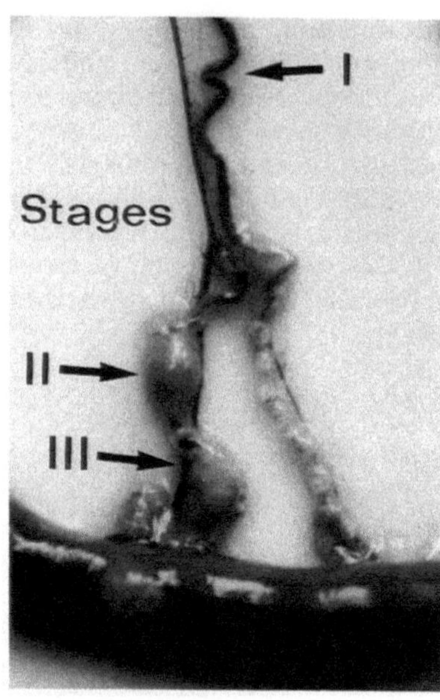

Fig. 8. Three stages of plaque formation in distal branches of the superior mesenteric artery of a 20-month-old SHR.

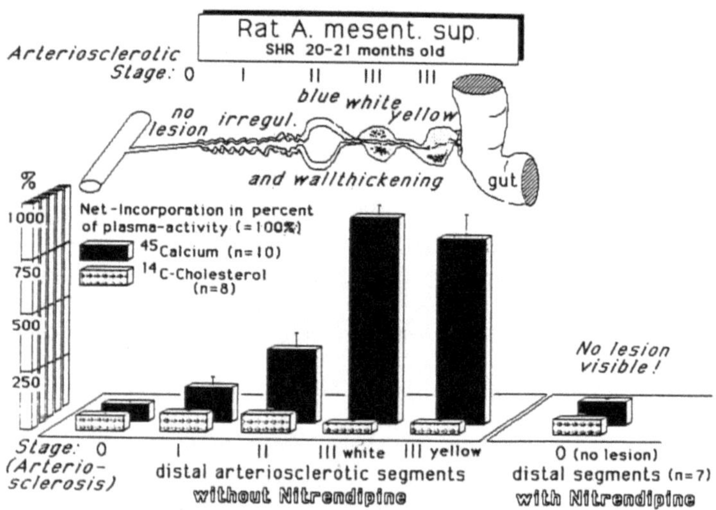

Fig. 9. Net in vivo incorporation of ^{45}Ca and ^{14}C-cholesterol into distal branches of 20- to 21-month-old SHRs (for classification of stages see Fig. 8).

arteriosclerotic alterations. Early lesions in arteries of SHRs were foci of Ca-overloaded, degenerated VSMCs with calcified mitochondria as well as Ca-loaded elastic fibres. Simultaneously, lipid-laden foam cells appeared in the intimal layer and cell proliferation started. Solid plaques (stage III) were characterized by an enormous wall thickening, migration, and massive cell proliferation, intracellular Ca overload, calcification of elastic fibers, discrete lipid accumulation, endothelial damage and thrombi as lethal complications (Fig. 10). Prophylactic long-term treatment with Ca antagonists, as for instance nitrendipine, completely prevented increase in blood pressure and plaque formation.

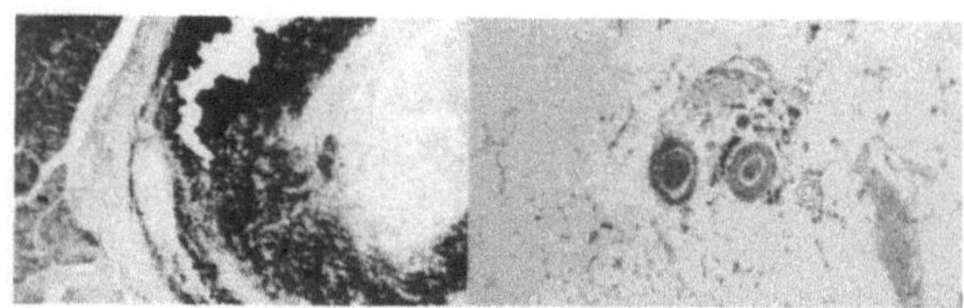

Fig. 10. Cross-sections of distal branches of superior mesenteric arteries of 21 month-old SHRs (v. Kossa's staining technique [1:100]). Ca is visualized as black deposits. Left: Fully developed plaque of stage III (see Figs. 8 and 9). Right: Antiarteriosclerotic protection by long-term treatment with nitrendipine.

2. Experimental arteriosclerosis in NaCl-fed Dahl-S rats

In 4-month-old NaCl-fed Dahl-S rats, a significant rise in total (209.2 ± 12.1 mg/dl [n = 27]) and LDL (79.4 ± 7.8 mg/dl [n = 10]) serum cholesterol was observed, compared to 23-month-old normotensive Wistar-Kyoto (total: 69.2 ± 4.7 [n = 10]); LDL: 16.97 ± 1.3 [n = 10]) and SHRs (total: 78.0 ± 4.4 [n = 10]); LDL: 16.46 ± 2.6 [n = 9]).

Arteriosclerotic lesions in Dahl-S rats were characterized by Ca plus cholesterol accumulation in the arterial walls, particularly in distal branches of the superior mesenteric arteries (Fig. 11). We treated Dahl-S rats simultaneously to an 8 % NaCl regimen with lovastatin (30–120 mg/kg b.w.) or with nitrendipine (10 mg/kg b.w.) for 4.5 months. The low dose of nitrendipine as well as lovastatin did not prevent blood pressure elevation (14-week-old Dahl-S rats: 203.0 ± 4.11 [n = 20] and 215.7 ± 4.5 mmHg [n = 21], respectively). Lovastatin significantly lowered serum cholesterol (total: 130.4 ± 8.4 [n = 16]; LDL: 26.5 ± 2.5 [n = 10]), nitrendipine did not (total: 243.0 ± 20.7 [n = 10]; LDL: 56.5 ± 8.3 [n = 10]). However, the Ca antagonist significantly reduced plaque formation in distal mesenteric branches of Dahl-S rats, compared to lovastatin (Fig. 12). The chemical microanalysis of arteriosclerotic plaques in Dahl-S rats confirmed these morphological data (Fig. 11). Lovastatin did not prevent Ca and cholesterol accumulation in the arterial walls. Low-dose nitrendipine significantly protected arteries from Ca overload and cholesterol incorporation.

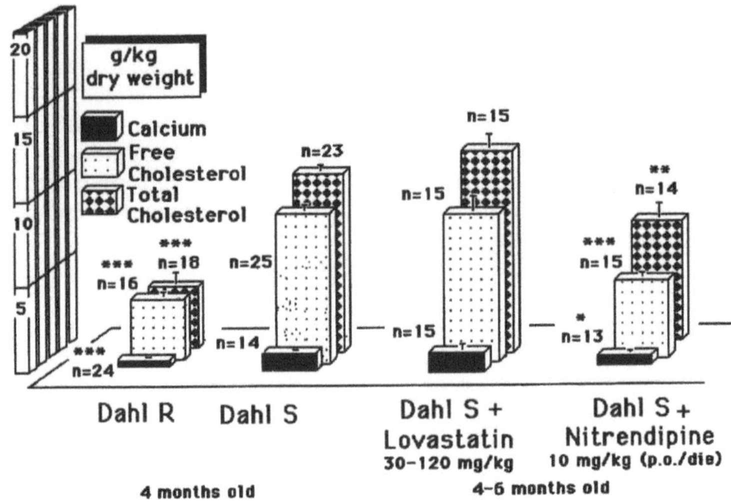

Fig. 11. Significant decrease of Ca and cholesterol contents in distal branches of superior mesenteric arteries of 8 % NaCl-fed Dahl-S-rats with nitrendipine (low dose), but not with lovastatin.

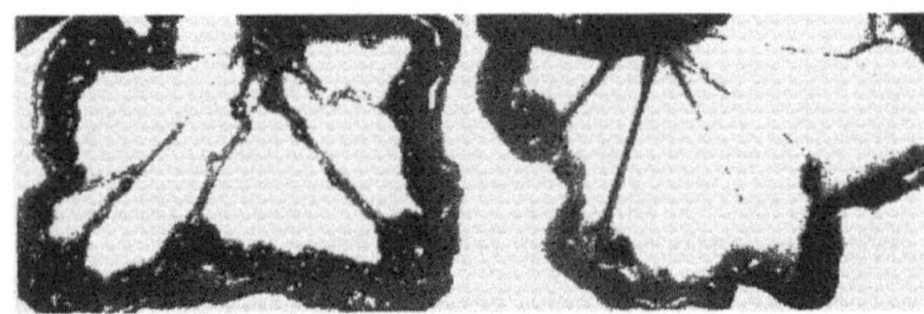

Fig. 12. Distal branches of superior mesenteric arteries of 4.5 month-old NaCl-fed Dahl-S-rats treated with lovastatin (left) and low-dose nitrendipine (right).

Discussion

Our microchemical analyses demonstrate that the development of conventional human coronary artery plaques is governed by a progressive Ca overload. Interestingly, already in early coronary lesions of type I ("fatty streaks") and II ("fibrous plaques"), arterial Ca incorporation was significantly increased compared to normal coronary segments of the same age groups. Conversely, only small amounts of free and total cholesterol were found without any quantitative correlation to plaque severity [9]. Apart from rare cases of familiar hypercholesterolemia [2], cholesterol accumulation per se can obviously not be responsible for coronary occlusion. However, oxidatively modified lipoproteins within the arterial wall may be decisively involved in plaque formation, possibly by making use of various messenger and "killer" functions of Ca ions [12]. Based on the widespread cholesterol hypothesis of

154

atherogenesis (2), diet and drug-induced lowering of serum cholesterol, particularly LDL, is propagated to prevent human arteriosclerosis. A possible interference with arterial Ca overload, however, is neglected to be a promising therapeutic principle (12). Two basic results, obtained from our experimental studies, deserve consideration.

1) It turns out that arterial Ca overload is a common denominator of all types of experimental arteriosclerosis or atheromatosis, tested so far (11, 14). But important differences existed with respect to the pathogenetic importance of Ca and Ca overload for plaque formation. In Ca-dominated (vitamin D_3-treated rats and rabbits) and mixed types (SHRs and NaCl-fed Dahl-S rats) of experimental arteriosclerosis an increased Ca uptake into the arterial walls already characterized the early lesions and promoted plaque formation. Arterial Ca overload in experimental atheromatosis of cholesterol-fed rabbits occurred at a late state of plaque formation and was indeed a secondary phenomenon, obviously not involved in coronary complications (14).

2) Ca antagonists (verapamil, nitrendipine, diltiazem) (6) specifically inhibited arterial Ca overload in all experimental models, presented above. In Ca-dominated and mixed types they obviously interfered with Ca-mediated early lesions and, consequently, stopped their deleterious sequelae within the arterial walls. In cholesterol-fed New Zealand rabbits, Ca antagonists inhibited the secondary Ca overload in aortae and coronary arteries. But they did not prevent intimal accumulation of foam cells and cholesterol, either in early lesions or in fully developed plaques (14). What are the Ca-consuming processes in arteriosclerosis and how may Ca antagonists interfere with arterial Ca overload?

Intracellular Ca overload

Ca ions – besides a multitude of vital biological functions – adopt the role of malignant killers, if cellular Ca metabolism runs out of control and intracellular Ca overload occurs (6, 10–12). Excess Ca signalling, overstimulation of Ca-dependent ATPases, lipases, proteases, endonucleases, and Ca-mediated mitochondrial damage are noxious consequences. Exhaustion of high-energy phosphates in concert with catabolism of phospholipid, protein and nucleic acids, and decrease of ionic pump activities induce cell death and necrosis (6, 12). Ca antagonists interfere with the initial step of this deleterious Ca cascade. By reducing transmembrane Ca supply they protect cells from cytotoxic Ca effects and maintain functional and morphological integrity. Ca-antagonistic cellular protection, first demonstrated for the mammalian myocardium (6), turned out to be a general principle regarding VSMCs (7, 8, 10–12), neurons (4), hepatic (22) and renal (5) tissues as well. In the Ca type of experimental arteriosclerosis, toxic Ca overload of VSMCs results from alteration of Ca homeostasis: The increase in extracellular Ca concentration up to 50 % by vitamin D_3 coincides with the nicotine-induced facilitation of transmembrane Ca uptake (11). In SHR several molecular mechanisms are discussed to be responsible for the tendency of VSMCs to Ca overload: Genetic dysregulation of voltage-dependent Ca channels (21), hyperactivity of membrane phospholipase C with subsequent liberation of IP3 and diacylglycerol (1), altered ATP-dependent calmodulin-stimulated Ca extrusion (18). In fact, leak, voltage- and receptor-mediated transmembrane Ca uptake into cultured native VSMCs were demonstrated to be increased in SHR compared to normotensive WKY rats (13). In arteries of cholesterol-fed New Zealand rabbits foci of calcified degenerated media cells were observed at a late stage – only in fully developed cholesterol plaques (14) at the intimal/media border. Oxidatively modified lipoproteins, shown to increase Ca transmembrane conductivity in VSMCs (26), may play a causative role. Suitable Ca antagonists were demonstrated to protect from Ca-induced media cell necrosis in all types of experimental arteriosclerosis and atheromatosis.

Extracellular Ca overload

Extracellular matrix macromolecules in arterial walls are known to possess high affinity-binding sites for Ca ions (19). Moreover, Ca ions were shown to potentiate cholesterol fixation at fibrous elastin (19). Ca-dependent elastase-type proteases, released from VSMCs, induce elastic fiber degradation; liberated elastin peptides exert chemotactic effects on VSMCs, increasing their migration towards the degradated elastin lamellae (16). Calcification and fragmentation of elastic fibers were associated with media cell necrosis in vitamin D_3-treated, spontaneously hypertensive, and NaCl-fed rats, as well as in cholesterol-fed rabbits. Suitable Ca antagonists completely prevented calcification and subsequent decay of elastic fibers in all experimental models. The underlying mechanism of action remains a matter of speculation: Ca antagonists might interfere with cellular processes involved in secretion, calcification and degradation of matrix proteins, or directly interact with extracellular Ca binding sites.

Messenger functions of Ca ions

Most of the cellular processes, involved in the development of conventional human arteriosclerosis, make use of messenger functions of Ca ions (17). In particular, Ca-modulated reactions of VSMCs are of central importance to all phases of plaque formation, irrespective of the exact sequence of events (20): Transformation to the synthetic type, migration, proliferation, synthesis and secretion of matrix proteins, to mention just a few. Similar events – especially migration and an enormous degree of cell proliferation simultaneously with intra- and extracellular Ca overload – were observed in distal branches of superior mesenteric arteries of SHRs and NaCl-fed Dahl-S rats. In SHRs, the progressive in vivo incorporation of ^{45}Ca into plaques of increasing severity indicated an exaggerated mural Ca metabolism, whereas only small amounts of cholesterol were accumulated. Hypertension and hypertensive arteriopathy may possess common pathogenetic roots: A defectively controlled Ca homeostasis in VSMCs (13, 24, 27). Arterial hypercontractility and the tendency to Ca overload are spectacular consequences (13). Moreover, Ca-modulated proliferation and even lipid accumulation were demonstrated to be augmented in cultured VSMCs of SHRs (27). Ca antagonists were shown to prevent the progressive Ca uptake into arterial walls of hypertensive rats. Consequently, they inhibited Ca-con-

156

suming processes in plaque formation. Probably by aiming at transmembrane Ca supply of VSMCs they may directly damp migration and proliferation (15) in addition to the above described protection from cell necrosis, as well as from calcification and fragmentation of elastic fibers.

In NaCl-fed Dahl-S rats, plaque formation is associated with a four-fold elevation of serum cholesterol, compared to Wistar rats and SHRs. Consequently, their mixed type of experimental arteriosclerosis, characterized by Ca plus cholesterol accumulations, represented a suitable model to study both the antiarteriosclerotic potencies of HMG Co reductase inhibitors – known to lower serum cholesterol – and of Ca antagonists – demonstrated to damp Ca-dependent events in lesion formation. Our results suggest that significant lowering of serum cholesterol with lovastatin does not prevent arterial cholesterol incorporation and plaque formation in Dahl-S rats. However, controlling of mural Ca metabolism with a low dose of nitrendipine, significantly reduced arterial Ca and cholesterol uptake as well as the development of arteriosclerotic alterations. This antiarteriosclerotic vasoprotection by nitrendipine was established in NaCl-fed Dahl-S rats independently of a concomitant antihypertensive effect of the drug and in the presence of hypercholesterolemia. Ca-dependent processes, sensitive to a Ca-antagonistic intervention, may play a decisive role in arterial lipid incorporation at least in NaCl-fed Dahl-S rats. In this context, messenger functions of Ca ions with respect to cellular cholesterol metabolism and Ca/cholesterol interactions at extracellular matrix molecules have to be considered (19, 27).

Conclusions

Killer and messenger functions of Ca ions were demonstrated to be decisively involved in the development of both Ca-dominated and mixed types of experimental arteriosclerosis, characterized by a progressive Ca overload. Using specific Ca antagonists as experimental tools, the prevention of an increased Ca uptake into arterial walls was demonstrated to be an effective antiarteriosclerotic principle in both experimental models. However, to what extent our experimental results can be transferred to human pathophysiology and, particularly, to human therapy, remains to be further clarified.

Acknowledgements. This paper is based on experimental studies supported by grants of the German Federal Ministry for Research and Technology.

The authors would like to express their gratitude for the secretarial work of Mrs. I. Reutenauer and for the continuous technical help of Mrs. L. Dumont, B. Seidel, and K. Zipfel.

References

1. Baudouin-Legros M, Meyer P (1990) Hypertension and Atherosclerosis. Journal of Cardiovascular Pharmacology 15 (Suppl 1): S1–S6
2. Brown MS, Goldstein JL (1984) How LDL receptors influence cholesterol and atherosclerosis. Sciences American 251 (5): 58–66
3. Czirfusz A, Matyas S, Fleckenstein-Grün G (1992) Reduction by calcium antagonists of nicotine-induced net calcium uptake into aortic media cells (Wistar rats). Pflügers Archiv 420 (Suppl 1): R 100
4. Dreyer EB, Kaiser PK, Offermann JT, Lipton ST (1990) HIV-1 coat protein neurotoxicity prevented by calcium channel antagonists. Science 248: 364–367
5. Epstein M (1991) Calcium antagonists and the kidney: Implications for renal protection. Journal of Cardiovascular Pharmacology 18 (Suppl 10): S64–S70
6. Fleckenstein A (1983) Calcium Antagonism in Heart and Smooth Muscle – Experimental Facts and Therapeutic Prospects; Monograph, John Wiley Publishing Company, New York Chichester Brisbane Toronto Singapore
7. Fleckenstein A, Fleckenstein-Grün G (1980) Cardiovascular protection by Ca antagonists. European Heart Journal 1: 15–21
8. Fleckenstein A, Fleckenstein-Grün G, Frey M, Zorn J (1987) Future directions in the use of calcium antagonists. American Journal of Cardiology 59: 177B–187B
9. Fleckenstein A, Frey M, Thimm F, Fleckenstein-Grün G (1990) Excessive mural calcium overload – A predominant causal factor in the development of stenosing coronary plaques in humans. Cardiovascular Drugs and Therapy 4: 1005–1014
10. Fleckenstein A, Frey M, Zorn J, Fleckenstein-Grün G (1987) The role of calcium in the pathogenesis of experimental arteriosclerosis. Trends in Pharmacological Sciences 8: 496–501
11. Fleckenstein A, Frey M, Zorn J, Fleckenstein-Grün G (1990) Calcium, a neglected key factor in hypertension and arteriosclerosis. Experimental vasoprotection with calcium antagonists or ACE inhibitors. In: Laragh JH, Brenner BM (eds) Hypertension: Pathophysiology, Diagnosis, and Management; Raven Press, New York, pp 471–509
12. Fleckenstein-Grün G, Fleckenstein A (1991) Calcium – A neglected key factor in arteriosclerosis. The pathogenetic role of arterial calcium overload and its prevention by calcium antagonists. Annals of Medicine 23: 589–599
13. Fleckenstein-Grün G, Frey M, Thimm F, Hofgärtner W, Fleckenstein A (1992) Calcium overload – An important cellular mechanism in hypertension and arteriosclerosis. Drugs 44 (Suppl 1): 23–30
14. Fleckenstein-Grün G, Thimm F, Frey M, Fleckenstein A (1993) Antiarteriosclerotic effects of calcium antagonists. Do calcium antagonists inhibit cholesterol accumulation in coronary arteries of cholesterol-fed rabbits? In: Godfraind T, Govoni S, Paoletti R, Vanhoutte PM (eds) Calcium Antagonists. Pharmacology and Clinical Research; Kluwer Academic Publishers, Dordrecht, pp 139–148
15. Henry P (1990) Calcium channel blockers and atherosclerosis. Journal of Cardiovascular Pharmacology 16 (Suppl 1): S12–S15
16. Hornebeck W, Robert L (1987) Transition of smooth muscle cells from the quiescent to the proliferative form. Role of elastonectin and plasma membrane elastase-type protease. Cardiovascular Disease 26: 219–225
17. Parmley WW, Blumlein S, Sievers R (1985) Modification of experimental atherosclerosis by calcium-channel blockers. American Journal of Cardiology 55: 165B–171B
18. Resink TJ, Bühler FR (1989) Dysfunctions of calcium extrusion in hypertension. In: Meyer P, Marche P (eds) Blood Cells and Arteries in Hypertension and Atherosclerosis; Raven Press, New York, pp 157–169
19. Robert L, Labat-Robert J, Hornebeck W (1986) Aging and atherosclerosis. Atherosclerosis Reviews 14: 144–170
20. Ross R (1986) The pathogenesis of atherosclerosis – an update. New England Journal of Medicine 314: 488–500

21. Rusch NJ, Hermsmeyer K (1988) Calcium currents are altered in the vascular muscle cell membrane of spontaneously hypertensive rats. Circulation Research 63: 997–1002

22. Seydewitz V, Staubesand J (1983) Calcifizierung der Arterienwand als Alternsprozeß. Elektronenmikroskopische und biochemische Untersuchung am Modell der Vitamin D_3-Überdosierung. Akta Gerontol 13: 114–118

23. Thurman RG, Apel E, Lemasters JJ (1988) Protective effect of nitrendipine against hypoxic injury in perfused livers from ethanol-treated rats. Journal of Cardiovascular Pharmacology 12: S113–S116

24. Triggle DJ (1990) Calcium: The control of a charismatic cation. In: Abraham s, Amitai G (eds) Calcium Channel Modulators in Heart and Smooth Muscle: Basic Mechanisms and Pharmacological Aspects; VCH, Weinheim/Deerfield Beach FL, and Balaban, Rehovot/ Philadelphia, pp 1–13

25. Virchow R (1856) Phlogose und Thrombose im Gefäß-System. Gesammelte Abhandlungen zur wissenschaftlichen Medizin. Meidinger Sohn u Co, Frankfurt

26. Weisser B, Locher R, Mengden T, Vetter W (1992) Oxidation of low density lipoprotein enhances its potential to increase intracellular free calcium concentration in vascular smooth muscle cells. Arteriosclerosis and Thrombosis 12: 231–236

27. Yamori Y, Nara Y, Shimizu S, Mano M, Nabika T (1989) Common cellular mechanisms in the development of hypertension and atherosclerosis. In: Meyer P, Marche P (eds) Blood Cells and Arteries in Hypertension and Atherosclerosis; Raven Press, New York, pp 233–245

Author's address:
Prof. Dr. med. G. Fleckenstein-Grün
Study Group for Calcium Antagonism
Physiological Institute of the
University of Freiburg
Hermann-Herder-Str. 7
D-79104 Freiburg
FRG

Role of calcium antagonists in progression of arteriosclerosis. Evidence from animal experiments and clinical experience.

Medizinische Universitätsklinik, Abteilung Innere Medizin III – Kardiologie und Angiologie – Ärztlicher Direktor Professor Dr. H. Just

Part I

Preventive effects of calcium antagonists in animal experiments*

M. Frey, H. Just

Summary: The quantitative predominance of free and total cholesterol over the amount of mural calcium is a most significant criterion of healthy human coronary arteries during the whole life span (0 – 90 years). However, this normal ratio increasingly changes as soon as arteriosclerotic alterations of the coronary walls set in. Accordingly, the mural calcium content steadily rises from fatty streaks over severe arteriosclerosis and, lastly, seems to reach a climax in plaques which caused lethal coronary infarction. Furthermore, the severe arteriosclerosis of human *art. dorsalis pedis* with gangrene (and amputation) is characterized by a tremendous calcium incorporation and absence of any mural cholesterol changes. Only in rare cases of human basilary plaques was a dangerous cholesterol incorporation in brain arterial wall found without significant elevation of serum cholesterol levels.The presented data indicate the existence of two different types of arteriosclerosis in one and the same patient and two basically different types of experimental coronary plaques according to their chemical composition, microscopic aspect and responsiveness to calcium antagonists: 1) the calcium type, developing in vitamin-D3-treated rats, and 2) the cholesterol type, represented by fatty coronary atheromata of cholesterol-fed rabbits. Coronary atheromata of cholesterol-fed New Zealand rabbits may be suitable models for coronary heart disease in rare cases of human familiar hypercholesterolemia. The formation of conventional human coronary artery plaques, however, essentially requires a progressive uptake of calcium, thereby representing a calcium dominated type of arteriosclerosis. Calcium antagonists specifically inhibit progredient mural calcium uptake in all experimental models of arteriosclerosis tested so far. However, neither in atheromatous arteries nor in afflicted organs (myocardium, liver, kidneys) of cholesterol-fed rabbits were we able to find any significant prevention of cholesterol accumulation by calcium antagonist.

Key words: Calcium antagonists – human arteriosclerosis – experimental arteriosclerosis – vasoprotection

Introduction

The current concept of the development of the arteriosclerosis is dominated by the assumption that the main etiological factor is the accumulation of lipids, particularly cholesterol, in the arterial walls. However, this notion neglects the fact that, in addition to cholesterol, high concentration of calcium salts, which also possess pathogenetic properties, are obviously another important constituent of sclerotic vessels. Max Bürger (4) implied in his observation in 1939 a parallelism between the

---

* Experimental work was done at the Physiological Institute of the University of Freiburg, study group for calcium antagonism, Head Prof. G. Fleckenstein-Grün

increase in cholesterol and calcium content of human aortae during an age span of
10 – 70 years. Despite this correlation, the main investigations have focused since
Anitschkow's cholesterol induced atheromathosis in rabbits (in 1919) almost exclusively on mural lipid accumulation in the search for the origin of arteriosclerosis (1).
Recent data from normal and arteriosclerotic human segments by Fleckenstein and
coworkers (7 – 35) have provided cumulative evidence that both augmentation of
arterial wall cholesterol and Ca are latent precursors of manifest arteriosclerosis of
senescent arteries. Moreover, calcium overload of the arterial vasculature is as
destructive as experimental cholesterol incrustation (7–35, 45, 48). Thus, the simultaneous arterial calcinosis cannot be considered a phenomenon of secondary importance.

The knowledge of the role of calcium in the development of human arterial
stenosing plaques probably broadens the indications for calcium antagonists to vasoprotective purposes, particularly in the elderly.

Methods

The calcium content was analyzed by atomic absorption spectrometry (Perkin Elmer,
Type 3030) after addition of lanthanum oxide to a final concentration of 0.5 % to
reduce interference with other minerals (as described in (24)). Wall cholesterol determination was done by gas chromatographic methods using a varian gas chromato-
graph 3400 equipped with a capillary column and following temperature conditions:
Injector 290 °C, column 280 °C, detector 290 °C, isothermal operating conditions.
The analysis of free cholesterol was performed after the lipid extract (in chloroform,
methanol, acetone) had been evaporated to dryness, redissolved and silylated with
MSTFA (n-methyl-n-trimethylsilyltrifluroacetamide). Cleavage of cholesterol esters
was done in an aliquot of the lipid extract after evaporation and redissolving in
methanol and HMDS (hexamethyl-disiloxane) at 65 °C over night. Subsequently, the
solution was again evaporated to dryness, redissolved in methanol and TMCS
(trimethyl-chlorosilane) and heated for 2 h at 65 °C. After evaporation MSTFA was
added and the total cholesterol content could be measured (24).

Results and Discussion

Calcium in the progression of human arteriosclerosis

The role of calcium in the development of human stenosing arterial plaques was
investigated by a direct approach, i.e., by careful chemical analysis of both healthy
human arterial walls and arteriosclerotic plaques of different severity. The natural
aging process of human arteries such as coronary, mesenteric, femoral, dorsalis pedis
and aortic wall tissue, is reflected in a marked augmentation of calcium content.
Spontaneous age-induced Ca accumulation is clearly a continuously developing process which starts many years before cytotoxic degrees of Ca overload are reached. At
the age of 60 – 70 years the situation becomes critical and the structural integrity of
the arterial walls is increasingly jeopardized. Figure 1 demonstrates that the mural
concentrations of calcium, free and total cholesterol in healthy human coronary

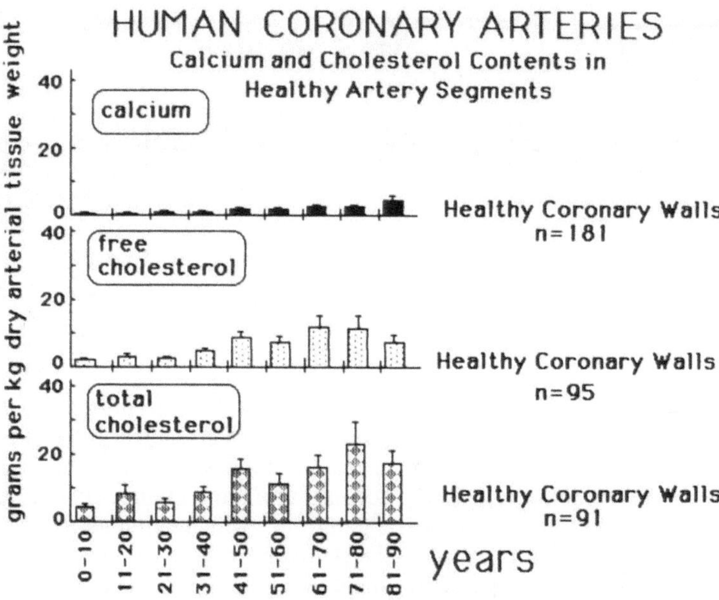

Fig. 1. Calcium and cholesterol content in healthy human coronary segments (age span 0 – 90 years).

arteries tend to rise with increasing age: For example, in the first age group of 0 – 10 years, 0.39 ± 0.04 (SEM) g calcium / kg dry arterial tissue weight was found, while at an age of 81 – 90 years, the calcium content amounted to 4.1 ± 1.0 g/kg dry weight. Similarly, the free and total cholesterol concentrations rose continuously. Thus, at an age of 0 – 10 years, about 2.31 ± 0.05 g free mural cholesterol and 4.65 ± 1.0 g total cholesterol / kg dry weight were measured. In the advanced age group of 61 – 90 years, the free and total cholesterol contents reached a maximum of about 12 and 20 g/kg, respectively, and the calcium contentration remained in a range of about 2.5 g/kg dry tissue weight. The quantitative predominance of free and total cholesterol over the amount of mural calcium is a most significant criterion of healthy coronary arteries during the whole life span.

The graph in Fig. 2b, however, demonstrates that in severe coronary plaques with fatal consequences the absolutely dominant constituent is calcium. On the other hand, the free and total cholesterol concentrations are only slightly higher than those in healthy coronary walls of the same age (Fig. 1). Thus, the natural predominance of cholesterol in healthy coronary segments becomes overshadowed by an excessive calcium incrustation in severe coronary plaques: The mural calcium content exceeds that of healthy coronary arteries by as much as 80 times. Similar imbalance of the natural ratio of calcium and cholesterol in favor of tremendous calcium incrustation can also be observed in the arteriosclerosis of human a. dorsalis pedis. The analytical data of calcium and cholesterol contents found during progressive arteriosclerosis of the human dorsal foot arteries are shown in Fig. 3a. Here, the development of arteriosclerosis usually starts at an age of 40 years, and is marked by a progressive calcium accumulation up to cytotoxic concentrations. As compared with the calcium

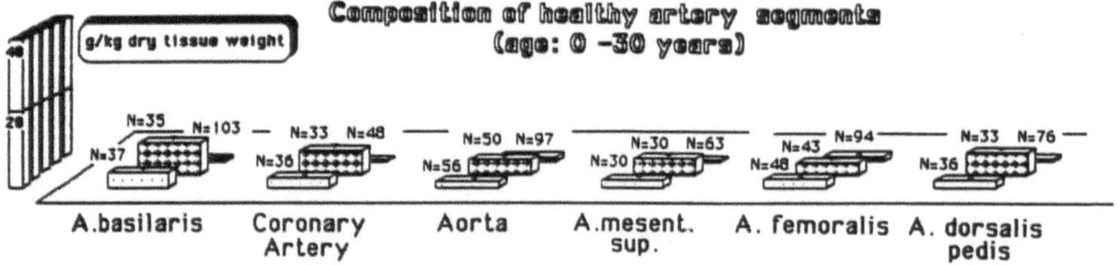

Fig. 2A. Chemical composition of healthy artery segments (age 0 – 30 years).

content of the dorsal foot arteries of children in the first decade, 81 to 90 year-old humans exhibited a rise in calcium concentrations by 53 times. In contrast, the old arteries do not contain more cholesterol than in the first decade of life: Any correlation between the tremendous calcium accumulation and changes in cholesterol content is absent. Furthermore, in diabetic gangrene, which can lead to amputation of the leg, the arteriosclerotic art. dorsalis pedis as well as art. tibialis anterior are characterized by a deleterious further increase in the calcium content, whereas the cholesterol content remains unchanged (Fig. 3b). Considering this higher calcium-

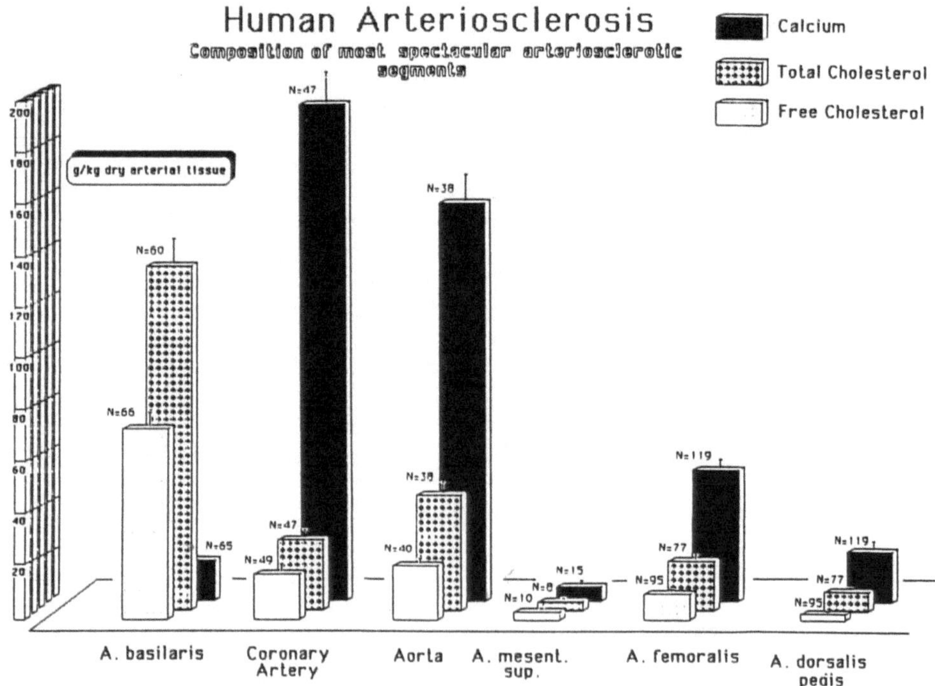

Fig. 2B. Chemical composition of human severe arteriosclerotic segments.

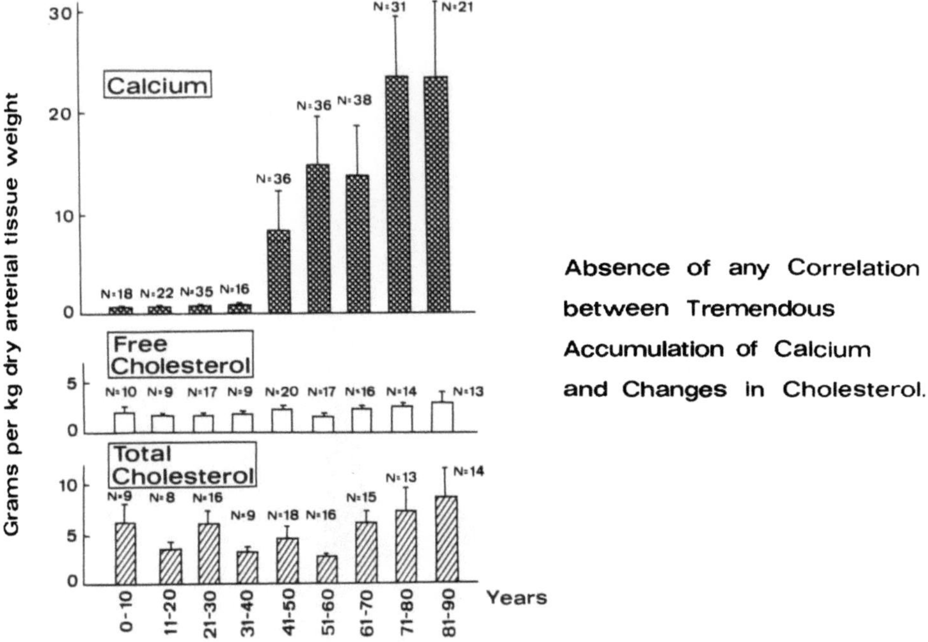

Fig. 3A. Age dependent arteriosclerosis of human arteria dorsalis pedis. Absence of any correlation between tremendous accumulation of calcium and changes in cholesterol.

affinity of aging a. dorsalis pedis in human diabetic gangrene, the efficacy of therapeutic cholesterol-lowering regimen to stop further arteriosclerotic progress in the leg remains dubious. Besides these calcium dominated types of human arteriosclerosis, i.e., in the coronary, femoral, mesenteric, tibialis ant., dorsalis pedis and aortic tissue, there exists a special colesterol-type which is represented by the basilar artery. As shown in Fig. 2 the mass of free and total cholesterol prevails in such arteriosclerotic basilar segments in comparison with calcium. However, these atheromata of brain vessels are less frequent than the development of coronary plaques. Whereas, on the one hand, in the chemical composition (calcium and cholesterol contents) of healthy, human, 0 – 30 year old arteries no difference exists between basilary, coronary, mesenteric, femoral, dorsalis pedis arteries or aortic tissue (see Fig. 2a), arteriosclerotic segments, on the other hand, from various sites in the same individual show their own characteristic arteriosclerotic fingerprints, so that after chemical analysis one can decide from which arterial site the arteriosclerotic plaques have arrived.

These analytical data show that human arteriosclerosis is characterized by an individual arteriosclerotic appearance of each vessel with the dominance of the calcium type of arteriosclerosis. Experimental models of arteriosclerosis in animals therefore have to consider both types of arteriosclerosis.

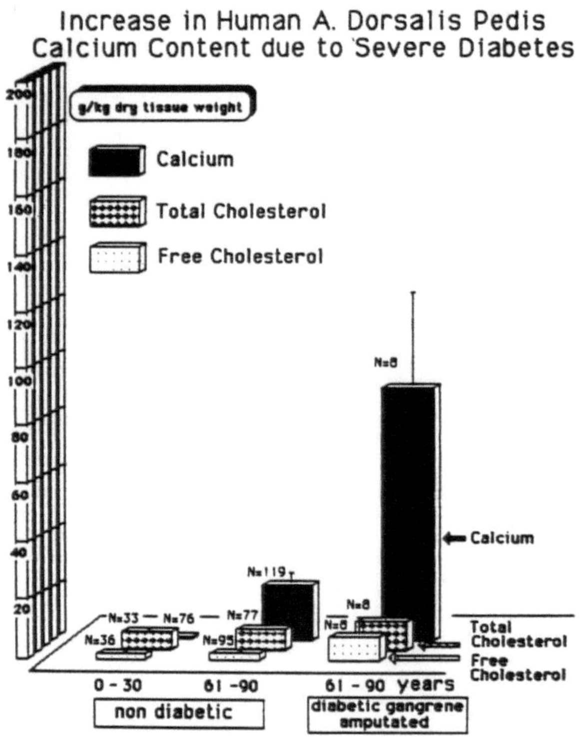

Fig. 3B. Acceleration of age-induced human arterial calcium overload by the risk factor diabetes mellitus.

Observations in animal experiments

A) Calcium-dominated type of arteriosclerosis: Table 1 summarizes experimental studies carried out since 1971 at the Physiological Institute in Freiburg (7–35). The best correspondence to the above-mentioned human calcium type of arteriosclerosis (represented by human a. dorsalis pedis and severe coronary artery plaques) was achieved by administration of high doses of vitamin D_3 to rats (6, 43). Under this condition calcium overload occurred within only a few days with simultaneous structural damage of the arterial walls as, for instance, of rat coronary, renal, mesenteric arteries and aortic tissues (41, 44). Interestingly, the basilar artery was spared from this deleterious calcium incrustation (35). This experimentally observed calcium resistance of rat brain arteries shows a parallelism to similar analytical data in human basilar artery, as mentioned above. Dangerous degrees of arterial calcium overload and morphological impairment resulted in the combination of vitamin D_3 plus nicotine (twice daily 25 mg p.o./kg rat body weight). This incompatible combination leads to severe arterial calcification which is so spectacular that spiral plaques can easily be recognized with simple optical aids in the rat descending branch of the left coronary artery (23). The calcium antagonist diltiazem, for instance, prevents this visible calcification (and lethal consequence) as shown in Fig. 4. Prophylactic antiar-

166

Table 1. Types of Arterial Calcinosis and Arteriosclerosis in Rats.

Types	Effective Ca Antagonists Examined
Intoxication with Vitamin D_3	Verapamil, Diltiazem
Hereditary Spontaneous Hypertension (Okamoto Rats; SHRs)	Nifedipine, Nisoldipine, Nitrendipine, Nimodipine, Amlodipine, Verapamil
NaCl-Sensitive Hypertension (Dahl-S Rats)	Nitrendipine, Anipamil, Felodipine, Amlodipine, Verapamil
Renal Goldblatt Hypertension	Nifedipine
Advanced Age	Verapamil
Alloxan Diabetes	Verapamil
Chronic Oral Doses of Nicotine	Nifedipine, Verapamil, Diltiazem
Combined Administration of Vitamin D_3 plus Nicotine	Diltiazem, Verapamil

teriosclerotic efficacy of magnesium and potassium salts were described by Selye 1958 (53) in dihydrotachysterol-intoxicated rats.

Age-dependent calcium accumulation clearly proceeds much faster in spontaneously hypertensive rats (SHRs) than in the non hypertensive controls. Figure 5 shows:

a) the slow natural age-dependent increase in the aortic and mesenteric calcium contents of normotensive Wistar-Kyoto control rats (WKY), and

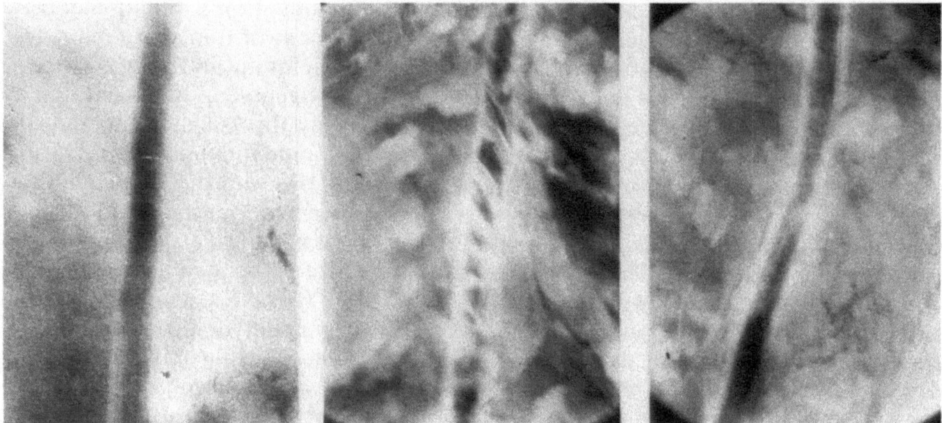

Fig. 4. Stereomicroscopic aspect of severe calcification of the rat coronary wall tissue with spiral calcified plaques in the descending branch of the left coronary stem artery by the incompatible drug combination of vitamin D_3 (300.000 IU/kg i.m.) and nicotine (2 × 25 mg/kg for 4 days p.o.). Diltiazem (2 × 50 mg/kg s.c. for 4 days) prevents visible calcification.

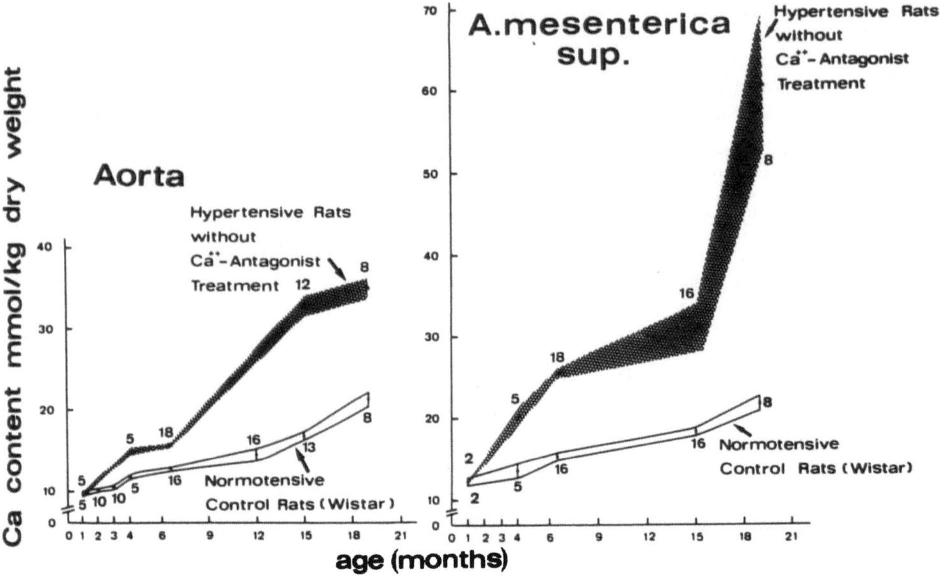

Fig. 5. Progressive age-dependent rise in Ca content of the aortic and mesenteric arterial walls of normotensive control wistar rats and of spontaneously hypertensive Okamoto rats (SHRs) without prophylactic calcium antagonistic treatment.

b) the much steeper rise of the arterial calcium incorporation that occurs in SHRs versus controls.

A chronic treatment of SHRs with the calcium antagonists verapamil, nifedipine, nitrendipine or nisoldipine in the food over a period of 20 months successfully inhibited arterial calcium overload and hypertension in parallel (38). Simultaneously, severe structural alterations (wall thickening, enlargement of transverse diameter, tortuosity, luminal obliterations, cholesterol depositions, intima and media wall proliferation, arteriitis nodosa and thrombosis) could be prevented as demonstrated in macroscopic and histological evaluations. Moreover, by this calcium antagonistic treatment the prolongation of life expectancy was a common finding (Fig. 6a). Similar results were obtained by long-term treatment of SHRs with the ACE-inhibitor captopril (23). In further experiments with salt-sensitive hypertensive Dahl rats the normalization of blood pressure by long-term treatment with the calcium antagonist nitrendipine was rewarded by a spectacular four-fold prolongation of life expectancy (Fig. 6b). Similar results were reported by Garthoff (37) and Kazda (46). Interestingly, long-term treatment with the antihypertensive agent dihydralazine in order to prevent vascular structural damage was a failure: Chronic lowering of blood pressure in SHRs was accompanied by a high mortality. Despite a normalized systolic blood pressure by dihydralazine arteriosclerotic lesions still occurred, indicating that blood pressure normalization by dihydralazine is no guarantee for vascular protection in animal experiments and, furthermore, signals a *split* in lowering arterial blood pressure and vascular protection. This discrepancy in blood pressure normalization and vascular protection was *never* observed in a calcium antagonistic long-term therapy.

168

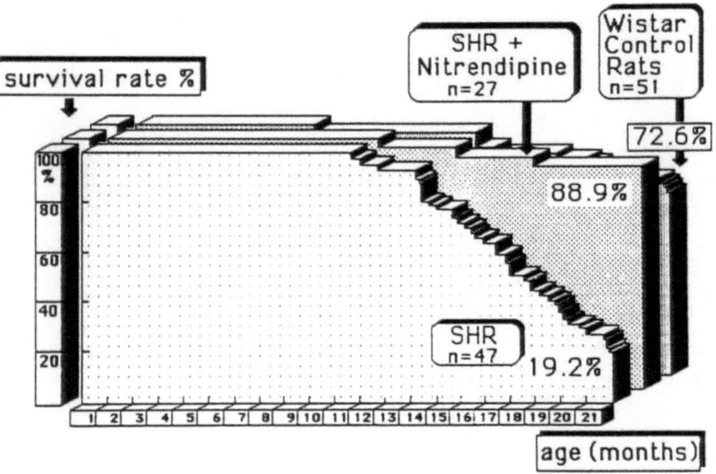

Fig. 6A. Prolongation of life expectancy of SHRs by long term treatment with the calcium antagonist nitrendipine.

Moreover, even low doses of calcium antagonistic therapy in salt-induced hypertensive Dahl rats *without* affecting high systolic blood pressure resulted in an impressive prolongation of life expectancy and reduction in vascular lesion. A sufficient vascular protection also of small vessels like arterioles by calcium antagonists can easily be observed in the rat ocular fundus as shown in Fig. 7. Untreated SHRs, salt-sensitive hypertensive Dahl rats or Goldblatt-hypertensive rats exhibit, in parallel to blood pressure elevation and duration of hypertension, a dramatic narrowing of retinal arterioles with tortuosity, aneurysm-like protuberances and, finally, retinal bleeding

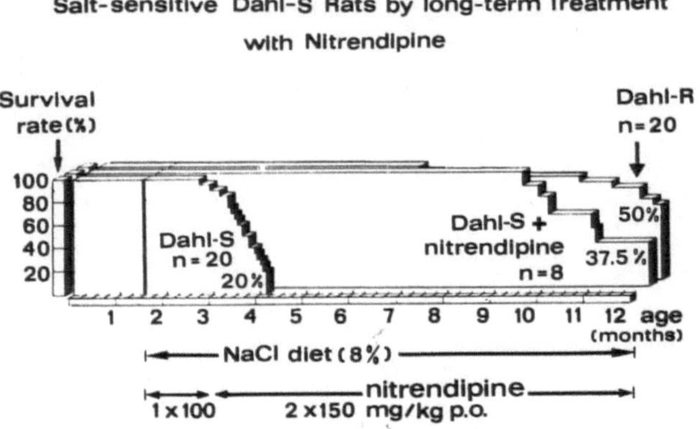

Fig. 6B. Normalization of natural life expectancy of saltsensitive Dahl rats by long-term treatment with nitrendipine.

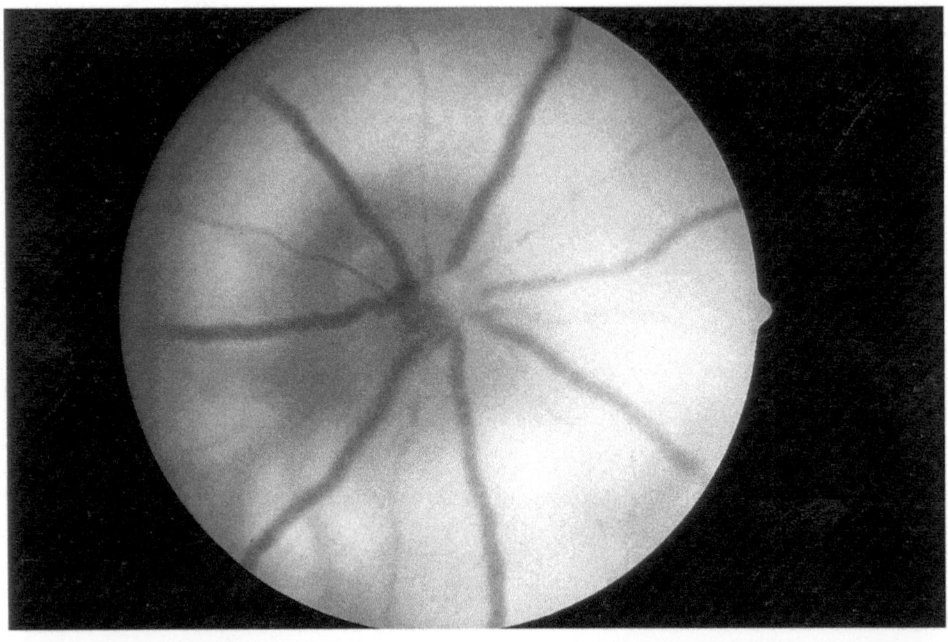

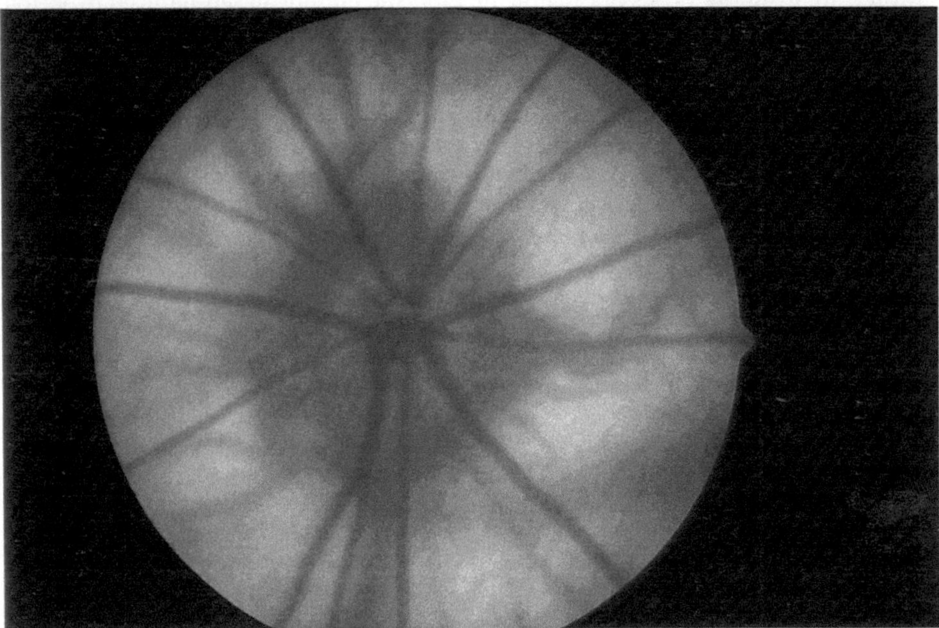

Fig. 7. Development of arteriosclerosis in the rat ocular fundus of an untreated hypertensive Dahl rat (upper picture) in contrast with the preservation of arteriolar wall integrity by nitrendipine (lower picture).

170

as demonstrated in the upper part in Fig. 7. The veins are normally unchanged. In our detailed measurements and calculations the reduction of retinal blood-perfusion in such hypertensive rats is dangerous and explains the pale aspect of those ocular fundi. Full arteriolar protection results from (lower part of Fig. 7) treatment with the calcium antagonist nitrendipine. The chemical analysis of calcium and cholesterol of rat mesenteric arteriosclerotic segments of untreated SHRs, hypertensive Dahl rats or hypertensive Goldblatt rats (23) corresponds to the above-mentioned data of calcium dominated human coronary plaques.

The natural calcium incorporation in aging arteries of *normotensive* rats is inhibited by a long-term calcium antagonistic treatment as shown in Fig. 8, indicating that this natural arterial calcinosis (as observed in humans, rats and dogs (11) is by no means an inevitable destiny (20, 23, 57, 58, 60). Prophylactic treatment with the calcium antagonist verapamil can indead interfere in rat experiments with the natural arterial senescence. As already pointed out, a prophylactic therapy with certain calcium antagonists (see Table 1) can also prevent the arteriosclerotic disaster provoked by the risk factors hypertension, diabetes mellitus, chronic nicotine administration or overdoses of vitamin D_3.

B) Cholesterol-dominated type of arteriosclerosis: Since Anitschkow introduced cholesterol-intoxicated rabbits in 1913 (1), their fatty arterial atheromata were proposed as models for human arteriosclerosis and, equally misleading, as targets for the experimental evaluation of arteriosclerotic drug potencies. But apart from rare cases of familial hypercholesterolemia, the formation of conventional human coronary

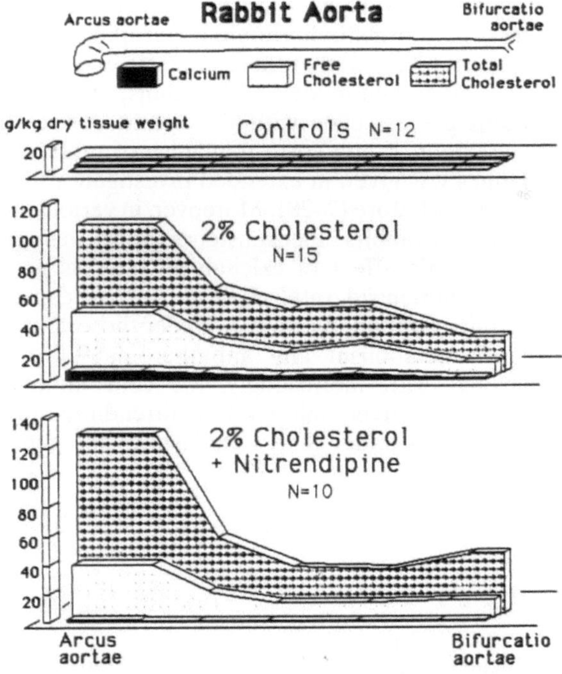

Fig. 8. Selectively inhibition of calcification in the atheromatotic aorta of cholesterol fed New Zealand rabbits by nitrendipine.

plaques, arteriosclerosis of the a. dorsalis pedis, and severe plaques of aortic tissue is governed by a progredient arterial calcium accumulation. On the other hand, ordinary human basilary plaques are the result of a severe cholesterol/calcium imbalance in favor of a massive cholesterol incrustation (see Fig. 2b). In our analytical investigations only in 8% of human basilary plaques could a significant calcium elevation be chemically verified, whereas the overwhelming part, i.e., 92% of basilary plaques, represented a cholesterol type of arteriosclerosis. So, cholesterol-intoxicated rabbits with severe aortic athermatosis may better represent a model for human basilary atheromatosis than for the calcium dominated type of human coronary plaques. Many studies in recent years were done to reduce the extension of atheromatotic lesions induced by a cholesterol-rich diet (1% or 2% cholesterol) fed to rabbits for 8 to 12 weeks. Henry and Bentely (42) first reported a suppression by nifedipine (20 mg twice daily) of atheromatosis in cholesterol-intoxicated rabbits besides an unchanged, abstruse high level of plasma cholesterol up to 2000 mg/dl. In this study aortic lesions covered 40% of the intimal surface, whereas only 15% of the aortic surface was covered in the nifedipine group. Similar vasoprotective results were obtained with verapamil (52, 3), nifedipine and nicardipine (59), isradipine (40), diltiazem (39, 55), flunarizine (2, 39, 50, 55), nisoldipine (36), nifedipine (49), and anipamil (5). But not all studies have been able to demonstrate that calcium antagonists prevent lesion progression in the cholesterol intoxicated rabbit model. Failure of nifedipine on atherogenesis in cholesterol-fed rabbits was reported by Stender et al. (54) and by Overturf (51). Moreover, the effects of calcium antagonists have been examined in Watanabe heritable hyperlipidemic rabbits – a genetic model of human homocygote familial hypercholesterolemia. Verapamil and nifedipine both failed to reduce atherosclerotic plaque formation in this model (39, 56). In an excellent overview (47) by Kroegh and Schroeder in 1991 concerning the failure of antiathertosclerotic effects of calcium antagonists in Watanabe heritable hyperlipidemic rabbits, the authors raised two important possibilities: "that there are differences in the pathogenesis of coronary disease between animal models and humans, and that atherogenesis may not be a homogeneous process even within the same species (namely rabbits)". Interestingly, this statement of non-correspondence of rabbit atheromathosis and human coronary plaques was given in extended investigations by Fleckenstein and coworkers a couple of years before (7–28). Moreover, in various series of cholesterol-induced atheromathosis in rabbits, Fleckenstein's group was not able to detect a significant antiatheromatotic effect of calcium antagonists given in a maximum dosage, which neither influenced total cholesterol food-consumption and body weight nor produced any obstipation (26–28). Under those controlled conditions of equivalent cholesterol metabolism any antiatheromatotic efficacy of calcium antagonists was lacking. But, interestingly, the concomitant calcinosis in the atheromatotic aorta was selectively inhibited by nitrendipine (Fig. 8), indicating a calcium antagonistic potency rather than a cholesterol antagonistic efficacy of calcium antagonists.

Conclusion

The present data indicate the existence of two basically different types of *human* and *experimental* arteriosclerotic plaques according to their chemical composition, microscopic aspect, and responsiveness to calcium antagonists:

1) The *calcium type,* developing in conventional human coronary plaques (with lethal consequences) and in human art. dorsalis pedis (with gangrene and amputation). This calcium type can experimentally be mimicked in vitamin-D_3-treated rats or in spontaneously hypertensive rats. Suitable calcium antagonists completely prevented this type of experimental arteriosclerosis. In our rat experiments a regression of arteriosclerotic lesions by long-term treatment with calcium antagonists could not be verified.

2) The *cholesterol type,* developing in human basilary plaques. This type of atheromatosis is represented by fatty atheromata of cholesterol-fed New Zealand rabbits. Intimal xanthomatosis following toxic hypercholesterolemia reflects the early lesions. Excessive cholesterol accumulation, resistant in our experiments to the prophylactic treatment with calcium antagonists, finally causes arterial occlusions. Though experimental animal models will never perfectly match human pathophysiology, both should have some characteristic features in common. Arterial atheromata of cholesterol-fed New Zealand rabbits may be suitable models for coronary heart disease in rare cases of human familiar hypercholesterolemia and in the conventional human basilary plaques.The formation of conventional human coronary artery plaques or severe arteriosclerosis of dorsal foot arteries, however, essentially requires a progressive uptake of calcium, thereby representing a calcium dominated type of arteriosclerosis. Calcium antagonists specifically inhibited progredient mural calcium uptake in all experimental models of arteriosclerosis tested so far.

Acknowledgement. Author MF is grateful to Prof. Dr. Fleckenstein-Grün, Head of study group for calcium antagonism at the Physiological Institute of Freiburg, for her creative supervision while I worked in that laboratory.This paper summarizes experimental work and analyses presented in several papers referenced herein. I am also grateful to Prof. Dr. H. Just for his creative and energetic activities.

References

1. Anitschkow N, Chalatow S (1913) Über experimentelle Cholesterinsteatose und ihre Bedeutung für die Entstehung einiger pathologischer Prozesse. Zentralblatt für Allgemeine Pathologie und Anatomie 14: 1–9
2. Betz E (1988) The effect of calcium antagonists on intimal cell proliferation in atherogenesis. Ann NY Acad Sci 522: 399–410
3. Blumlein SL, Sievers R, Kidd P et al. (1984) Mechanism of protection from atherosclerosis by verapamil in the cholesterolfed rabbit. Am J Cardiol 54: 884–889
4. Bürger M (1939) Die chemischen Altersveränderungen an Gefäßen. Z Neurol Psychol 167: 273–280
5. Catapono AL, Maggi FM, Cicerano U (1988) The antiatherosclerotic effect of anipamil in cholesterol-fed rabbits. Ann NY Acad Sci 522: 519–521
6. Eisenstein R, Zeruolis L (1964) Vitamin-D-induced aortic calcification. Arch Pathol 77: 27–35
7. Fleckenstein A (1983) Calcium antagonism in heart and smooth muscle – experimental facts and therapeutic prospects. Monograph, New York: John Wiley
8. Fleckenstein A (1987) Model experiments on anticalcinotic and antiarteriosclerotic arterial protection with calcium antagonists. J Mol Cell Cardiol 19 (Suppl II): 109–121

9. Fleckenstein A, Fleckenstein-Grün G, Janke J, Frey M (1977) Kardiovasculäre Grundwirkungen der Ca-Antagonisten und deren therapeutischer Wert in Kombination mit Herzglykosiden. Z präkl Geriatrie 7: 269–284

10. Fleckenstein A, Fleckenstein-Grün G (1980) Cardiovascular protection by calcium antagonists. Eur Heart J 1 (Suppl B) 15–21

11. Fleckenstein A, Frey M, Fleckenstein-Grün G (1983) Protection by calcium antagonists against experimental arterial calcinosis, In: Pyörälä K et al. (eds) Secondary prevention of coronary heart disease. Workshop of the Internat Soc and Federation of Cardiology, Titisee Oct 1983. Stuttgart: G Thieme Verlag 109–122

12. Fleckenstein A, Frey M, Leder O (1983) Prevention by calcium antagonists of arterial calcinosis. In Fleckenstein A et al. (eds) Drug development and evaluation – New calcium antagonists – Recent development and prospects. Diltiazem Workshop, Freiburg/Germany May 1982, Stuttgart: Gustav Fischer Verlag 15–31

13. Fleckenstein A, Frey M, v Witzleben H (1983) Vascular calcium overload – A pathogenic factor in arteriosclerosis and its neutralisation by calcium antagonists. In: Kaltenbach M, Neufeld HN (eds) New therapy of ischemic heart disease and hypertension. Proceed of the 5th Internat. Adalat Symp, Berlin May 1982, Amsterdam: Excerpta Medica 36–52

14. Fleckenstein A, Frey M, Fleckenstein-Grün G (1984) Cellular injury by cytosolic calcium overload and its prevention by calcium antagonists – a new principle of tissue protection. In Mechanisms of Hepatocyte Injury and Death, Proceed of the 38th Falk Symp, Basel Oct 1983, Keppler D, Popper H, Bianchi L, Reutter (eds) Kluwer Academic Publ Group, Lancaster, Boston, The Hague, Dordrecht: 321–335

15. Fleckenstein A, Frey M, Zorn J, Fleckenstein-Grün G (1985) Interdependence of antihypertensive, anticalcintic and antiarteriosclerotic effects of calcium antagonists – model experiments on spontaneously hypertensive rats. In: Fleckenstein A, van Breemen C, Gross R, Hoffmeister F (eds) Cardiovascular effects of dihydropyridine calcium antagonists and agonists. Bayer Symposium IX, Berlin: Springer Verlag 480–499

16. Fleckenstein A, Frey M, Fleckenstein-Grün G (1986) Antihypertensive and arterial anticalcinotic effects of calcium antagonists. Am J Cardiol 57: 1D–10D

17. Fleckenstein A, Frey M, Zorn J, Fleckenstein-Grün G (1985) Experimental basis of the long-term therapy of arterial hypertension with calcium antagonists. Am J Cardiol 56: 3H–14H

18. Fleckenstein A, Frey M, Fleckenstein-Grün G (1986) Zwanzig Jahre Calcium-Antagonismus aus physiologischer und pathophysiologischer Sicht, Rückblick und Ausblick. Österreichische Apotheker-Zeitung 40: 397–404

19. Fleckenstein A, Fleckenstein-Grün G, Frey M, Zorn J (1987) Future directions in the use of calcium antagonists. Am J Cardiol 59: 177B–187B

20. Fleckenstein A, Frey M, Zorn J, Fleckenstein-Grün G (1987) The role of calcium in the pathogenesis of experimental arteriosclerosis. Trends Pharmacol Sci (TIPS) 8: 496–501

21. Fleckenstein A, Fleckenstein-Grün G, Frey M, Zorn J (1989) Calcium antagonism and ACE inhibition. Two outstanding effective means of interference with cardiovascular calcium overload, high blood pressure, and arteriosclerosis in spontaneously hypertensive rats. Am J Hypertens 2: 194–204

22. Fleckenstein A, Fleckenstein-Grün G, Frey M, Thimm F (1990) Experimental antiarteriosclerotic Effect of calcium antagonists. J Clin Pharmacol 30: 151–154

23. Fleckenstein A, Frey M, Zorn J, Fleckenstein-Grün G (1990) Calcium, a neglected key factor in hypertension and arteriosclerosis. Experimental vasoprotection with calcium antagonists or ACE inhibitors. In Laragh JH, Brenner BM (eds) Hypertension: Pathophysiology, diagnosis and management, New York: Raven Press: 471–509

24. Fleckenstein A, Frey M, Thimm F, Fleckenstein-Grün G (1990) Excessive mural calcium overload – a predominant causal factor in the development of stenosing coronary plaques in humans. Cardiovascular Drugs and Therapy 4: 1005–1014

25. Fleckenstein-Grün G, Frey M, Fleckenstein A (1984) Calcium antagonists: Mechanisms and therapeutic uses. Trends in Pharmacological Sciences (TIPS) 5: 283–286

26. Fleckenstein-Grün G, Fleckenstein A (1991) Calcium – a neglected key factor in arteriosclerosis. The Pathogenetic role of arterial calcium overload and its prevention by calcium antagonists. Annals of medicine 23: 589–599

27. Fleckenstein-Grün G, Frey M, Thimm F, Luley C, Czirfusz A, Fleckenstein A (1991) Differentiation between calcium- and cholesterol-dominated types of arteriosclerotic lesions: Antiarteriosclerotic aspects of calcium antagonists. J Cardiovasc Pharmacol 18 (Suppl 6): S1–S9

28. Fleckenstein-Grün G, Thimm F, Frey M, Fleckenstein A (1993) Anti-arteriosclerotic effects of calcium antagonists: Do calcium antagonists inhibit cholesterol accumulation in coronary arteries of cholesterol-fed rabbits? In: Godfraind T et al. (eds) Calcium Antagonists. Kluwer Academic Publishers and Fondazione Giovanni Lorenzini, Netherlands: pp 139–148

29. Frey M, Janke J, Fleckenstein (1978) Inhibition of vitamin D3-induced vascular calcification in rats by means of the ca antagonist verapamil. Pflügers Arch Eur J Physiol 377: R9

30. Frey M, Keidel J, Fleckenstein A (1980) Verhütung experimenteller Gefäßverkalkungen (Mönckeberg's Typ der Arteriosklerose) durch Calcium-Antagonisten bei Ratten. In: Fleckenstein A, Roskamm H (eds): Calcium Antagonismus. Proceed Internat Symp in Frankfurt 1978, Berlin: Springer Verlag: 258–264

31. Frey M, v Witzleben H, Keidel J, Fleckenstein A (1980) Restriction of Ca overload of the arterial walls of spontaneously hypertensive rats (SHR) by Ca antagonists (verapamil, nifedipine). Naunyn-Schmiedeberg's Arch Pharmacol 313, R48

32. Frey M, Zorn J, Fleckenstein A, Fleckenstein-Grün G (1985) Normalization of high blood pressure, arterial calcification and mortality in spontaneously hypertensive rats (SHR) by long-term calcium antagonist treatment. Hochdruck 6: 29–32

33. Frey M, Zorn J, Fleckenstein A, Fleckenstein-Grün G (1986) Prevention of different types of experimental calcinosis of rat arteries by means of calcium antagonists. In: What is new in angiology? Trends and Controversies. Proc. 14th World Congress Internat Union of Angiology in Munich, Maurer PC et al. (eds) Zuckschwerdt-Verlag München–Berlin–Wien: 93–95

34. Frey M, Zorn J, Fleckenstein A, Fleckenstein-Grün G (1988) Protection of arterial and arteriolar wall structure by specific calcium antagonists. Annals of NY Academy of Sciences 522: 420–432

35. Frey M, Thimm F, Fleckenstein-Grün G, Fleckenstein A (1990) Peculiarities of vitamin D3-induced excessive calcium accumulation in different rat arteries. Pflügers Arch Eur J Physiol 415, R 64

36. Fronek K (1987) Effect of nisoldipine on diet-induced atherosclerosis in rabbit. Ann NY Acad Sci 522: 525–526

37. Garthoff B, Kazda S (1981) Calcium antagonist nifedipine normalizes high blood pressure and prevents mortality in saltloaded DS substrain of Dahl rats. Eur J Pharmacol 74: 111–112

38. Gasser R, Frey M, Byon YK, Fleckenstein-Grün G (1988) Restriction by nitrendipine of excessive concentration of free intravellular calcium ions in ventricular myocardium of hypertensive rats. Angiology 39: 246–252

39. Ginsburg R, Davis K, Bristow M et al. (1983) Calcium antagonists suppress atherogenesis in aorta but not in the intramural coronary arteries of cholesterol-fed rabbits. Lab Invest 49: 154–158

40. Habib JB, Bossaler C, Wells S et al. (1986) Preservation of endothelium-dependent vascular relaxation in cholesterol-fed rabbit by treatment with the calcium blocker PN 200–220. Circ Res 58: 305–309

41. Hass GM, Trueheart RE, Hemmens A (1960) Experimental arteriosclerosis due to hypervitaminosis D. Am J Pathol 37: 521–539

42. Henry PD, Bentley KI (1981) Suppression of atherogenesis in cholesterol-fed rabbits treated with nifedipine. J Clin Invest 68: 1366–1369

43. Herzenberg H (1929) Studien über die Wirkungsweise des bestrahlten Ergosterins (Vigantol) und die Beziehung der von ihm gesetzten Veränderungen zur Arteriosklerose. Beitr Path Anat 82: 27–56

44. Janke J, Hein B, Pachinger O, Leder O, Fleckenstein A (1972) Hemmung arteriosklerotischer Gefäßprozesse durch prophylaktische Behandlung mit MgCl2, KCl und organischen Ca-Antagonisten (quantitative) Studien mit 45 Ca bei Ratten). In: Betz E (ed) Vascular smooth muscle, Verh Satel Symp XXV Internat Congr Physiol Wissenschaften, Tübingen 1971. Berlin Springer Verlag, 71–72
45. Katase A (1913) Experimentelle Verkalkungen an gesunden Tieren. Beiträge Path Anat und Allg Path 57: 516–550
46. Kazda S, Garthoff B, Luckhaus G, Nash G (1983) The calcium antagonist nifedipine and its analogues preserve tissue integrity and life span in experimental malignant hypertension. In: Hashimoto K, Kawai C (eds) New Therapy of Ischemic Heart Disease and Hypertension. Asian Pacific Adalat Symposium in Tokyo 1982. Tokyo: Medical Tribune Inc 50–62
47. Keogh AM, Schroeder JS (1991) The antiatherogenic effects of calcium antagonists. Am J Hypertension 4: 512S–518S
48. Ketelsen UP (1989) Experimentelle Gefäßkalzinosen und Kalziumantagonisten. In: Gerald Klose (ed) Arteriosklerose, Molekulare und zelluläre Mechanismen, Sicherheit von Prävention und Therapie. Springer Verlag Berlin, Heidelberg New York London Paris Tokyo 39–46
49. Nayler WG (1991) Molecular mechanisms involved in the antiatherogenic effect of the calcium antagonists. In: Nayler WG (ed) The second generation of calcium antagonists. Springer Verlag Berlin, 139–151
50. Niekerk van JLM, Hendriks TH, De Boer HHM et al. (1984) Does nifedipine suppress atherogenesis in WHHL rabbits? Atherosclerosis 53: 91–98
51. Overturf ML, Smith SA (1986) Failure of nifedipine to reduce atherogenesis in cholesterol-fed Rabbits. Artery 13: 267–282
52. Rouleau JL, Parmley WW, Stevens J et al. (1983) Verapamil suppresses atherosclerosis in cholesterol-fed rabbits. J Am Coll Cardiol 1: 1453–1460
53. Selye H (1958) Prophylactic treatment of an experimental arteriosclerosis with magnesium and potassium salts. Am Heart J 55: 805–809
54. Stender S, Stender I, Nordestgaard B et al. (1984) No effect of nifedipine on atherogenesis in cholesterol-fed rabbits. Arteriosclerosis 4: 389–394
55. Sugano M, Nakitshima Y, Matsushima T et al. (1986) Suppression of atherosclerosis in cholesterol-fed rabbits by diltiazem injection. Arteriosclerosis 6: 237–241
56. Watanabe N, Ishikawa Y, Okamoto R et al. (1987) Nifedipine suppresses atherosclerosis in cholesterol-fed rabbits but not in Watanabe heritable hyperlipidemic rabbits. Artery 14: 283–294
57. Weinstein DB, Heider JG (1987) Antiatherogenic properties of calcium antagonists. Am J Cardiol 59: 163B–172B
58. Weinstein DB (1990) Protective effects of calcium channel antagonists in experimental models of atherogenesis and vascular disease. In: Laragh JH, Brenner BM (eds) Hypertension: Pathophysiology, diagnosis and management. New York: Raven Press 511–519
59. Willis AL, Nagel B, Churchill V (1985) Antiatherosclerotic effects of nicardipine and nifedipine in cholesterol-fed rabbits. Arteriosclerosis 5: 250–255
60. Yogamundi Moon J (1972) Factors affecting arterial calcification associated with atherosclerosis. A review. Atherosclerosis 15: 119–126

Authors' address:
Dr. M. Frey
Medizinische Universitätsklinik
Hugstetter Str. 55
D-79106 Freiburg
FRG

Part II
Clinical experience with preventive effects of calcium channel blockers in atheromatous coronary artery disease

H. Just and M. Frey

Summary: Experimental evidence for antiatheromatous of effects of calcium antagonists has been impressive. Clinical experience has, in contrast, been more difficult to obtain. Primary prevention with calcium antagonists has not been studied due to obvious difficulties. Secondary prevention, however, has been investigated:

Several studies have addressed influence of calcium antagonists upon atheromatous arterial wall changes as demonstrated by quantitative coronary angiocardiography. A review of these studies reveals considerable methodological problems. For nifedipine it could be demonstrated, however, that the occurrence of "new lesions" can be retarded to a certain extent (3–6). Nicardipine has been studied, but the preventive effect reported cannot be considered valid, because distribution of risk factors to the study groups was not statistically homogeneous.

Another approach has been the application of calcium antagonists to patients with acute myocardial infarction. Here, vascular and myocardial effects come into play.

In non-Q-wave, i.e. not transmural infarction, the calcium antagonist diltiazem definitely has preventive effects as regards re-infarction (2). The large multicenter post-myocardial infarction trial MDPIT showed an improvement of cardiac envent rate and re-infarction. This effect was seen only if pulmonary congestion was not present. Calcium antagonists have negative effects if cardiac failure is present.

Verapamil was shown to have beneficial effects in acute myocardial infarction in the large DAVIT trials. Here again, the effect was only seen if heart failure was not present. Otherwise negative results were recorded.

Nifedipine demonstrated only borderline myocardial protective effects in acute myocardial infarction (7).

We conclude that calcium antagonists have vascular and myocardial protective effects. These can be offset in the presence of cardiac failure after myocardial infarction. A definite, although quantitatively limited vasoprotective action has been shown for nifedipine and diltiazem.

Key words: Calcium antagonists − antiatheromatous effect − coronary arteries − myocardial infarction – nifedipine − nicardipine − diltiazem − verapamil

In clinical medicine primary prevention of coronary atheromatous disease is usually not feasable. Such an approach could only be tested in long-term studies in patients with known risk profile, beginning early before any clinical manifestation of the disease can be detected. This has not been possible so far and such studies have not been published. Instead secondary prevention has been taken into focus. Besides calcium channel blockers numerous agents have been tested for secondary preventive effects in acute and chronic coronary artery disease: Nitrates, beta-blockers, antiaggregatory and anticoagulating agents, antilipidemic agents and ACE-inhibitors. In this context, we shall, however, confine our considerations to the calcium channel blockers. It will have to be tested if and to what extent the overwhelming experimental evidence of an antiarteriosclerotic effect of the calcium antagonists, as summarized by G. Fleckenstein-Grün in the previous chapter and by us in the first part of this contribution, can be demonstrated in the clinical situation.

Prevention could result in
1) abolition of symptoms or signs;
2) reduction in incidence of complications of the disease, such as infarct or reinfarction, occurrence of unstable angina or sudden cardiac death.
3) Inhibition or slowing of the atheromatous arterial changes. These could manifest themselves either in a regression of manifest plaques with reduction of luminal narrowing or in prevention of the appearance of new plaques and/or stenoses.

It is exceedingly difficult to identify in a given clinical situation over a longer period of time changes of the status and structure of the arterial wall. Today, the only tool available is selective coronary angiography, especially if the vessel size can be assessed quantitatively. However, this presents, at best, the inner contour of the vessel. Direct information as to the size and thickness of the arterial wall is not available. The newer tool of intravascular ultrasound has not been used for the measurement of plaque and vessel wall changes over time. Direct histopathological assessment of the vascular wall has for obvious reasons not been possible so far. Therefore, we have to content ourselves with angiographic data. In view of strong experimental evidence for a direct myocardial protective effect of the calcium channel blockers, we shall also discuss the clinical evidence for the effectiveness of calcium channel blockers in acute myocardial infarction and in the long-term post-infarction period.

Symptomatic relief in coronary disease of effectiveness of the application of calcium channel blockers will have to be excluded from our considerations: The overwhelming experimental and clinical evidence for relaxation of vascular smooth muscle with improvement of the regional and global coronary circulation by all calcium channel blockers is dominant in these syndromes, thereby conceiling any evidence for a direct antiatherosclerotic effect on the vessel wall. Furthermore, it deserves mention that the pronounced blood-pressure-lowering effect of calcium channel blockers with normalization of arterial pressure and systemic vascular resistance in hypertensive patients will induce another confounding factor, if the antiatherosclerotic effect is to be defined.

Calcium channel blockers have numerous effects on the tissues, five of which can be considered operative in the prevention of atheromatosis:

1) prevention of Ca-overload;
2) upregulation of LDL-receptors;
3) inhibition of cell migration;
4) inhibition of SMC-proliferation;
5) antiplatelet effects.

How do these effects translate into clinical inhibition of progression or prevention of new lesions?

Loaldi and coworkers (6) published in 1989 a double blind randomized-three-arm study comparing nifedipine, propanolol, and isosorbid-dinitrate in patients with chronic coronary artery disease utilizing quantitative angiocardiography (Table 1). They gave 80 mg nifedipine per day over 2 years. The results, given in Table 1, show that progression was comparable in the three groups. However, the appearance of new lesions was only 10% in the nifedipine group and thereby statistically significantly lower (-77%) than in the two other groups. The study was well conducted; however, the angiocardiographic technique utilized only single plane filming and identical projections at the different times of study were not convincingly demonstrated in all instances.

Gottlieb and coworkers (4) published in 1989 a 1-year study with 60 mg nifedipine, again utilizing quantitative angiocardiography. Rather than native coro-

Table 1. DB R 3-arm CCAD QACG.

	Nifedipine	Propranolol	ISDN
n =	39	36	38
Progression	31 %	53 %	47 %
New lesions	10 %	34 %	29 %

Loaldi et al. AJC, 1989
DB R 3-arm: double-blind, randomized 3-arm design
CCAD: chronic coronary artery disease
QACG: quatitative angiocardiography

nary arteries, they studied aorto-coronary venous bypass grafts. Over the study period, they found a reduction in the appearance of new lesions by 34 % in the verum group as compared to placebo. This was statistically significant. However, the cases were relatively few and a longer observation period would have probably led to more convincing results.

In 1990, Lichtlen et al. (5) published the INTACT study (Table 2). In this multicenter, double blind, placebo-controlled, randomized trial in patients with chronic coronary artery disease, central, blinded, computer-assisted evaluation and measurement of the angiocardiographic films were done. The patients received 80 mg of nifedipine over 3 years. The number of patients participating allowed a statistically meaningful evaluation. Here again, progression of existing lesions was not influenced by the calcium channel blocker (Table 2). Spontaneous regression likewise was not different in the verum- and the placebo-group. On the whole, 2/3 of the patients showed no change in their coronary arteriographic stenosis pattern. However, the appearance of new lesions, well in keeping with the aforementioned studies, was reduced under nifedipine treatment: New lesions, i.e., wall irregularities detected by angiocardiography of more than 20 % luminal obstruction was observed in 144 instances under placebo and 103 × under nifedipine. This calculates to 0.82 new lesions per patient under placebo and to 0.59 new lesions per patient under nifedipine (Table 2). This difference of 28 % reaches a statistically significant differ-

Table 2. DB PC R CCAD QACG 3 years.

	Nifedipine	Placebo	
n =	173	175	
Progression	28 %	24 %	
Regression	10 %	14 %	
no change	63 %	65 %	
New lesions per patient	0.59	0.82	$p = 0.03$

INTACT Lichtlen, 1990
DB PC R: double-blind, placebo controlled, randomized design
CCAD: chronic coronary artery disease
QACG: quantitative angiocardiography

ence with a *p*-value of 0.03. The number of patients with new lesions as well as all parameters describing the size of existing lesions and their change over time did not show statistically significant differences. This well conducted study has been carefully evaluated and the reduction of appearance of new lesions under nifedipine can be considered valid, although the effect seems relatively small, smaller, than in the aforementioned studies.

Waters and coworkers (16) studied nicardipine in comparison to placebo (Table 3). This double blind, placebo controlled, randomized trial extended over 13 months. Patients with chronic coronary artery heart disease were studied with quantitative coronary angiography. Statistical comparison of the two groups was (as in the INTACT study) statistically valid. Except for the fact that here the number of those patients who continued to smoke during the study was statistically higher (p = 0.02) in the placebo group as compared to verum. Also, a conspicuous difference in age distribution is found, in that the patients receiving nicardipine were slightly older than those on placebo.

Table 3 shows that, here again, progression and regression of existing lesions were not different in the two groups. Statistically significant differences were found in progression per patient and progression per lesion in minimal lesions. Although the definition for the appearance of new lesions is similar as in the INTACT trial or the Loaldi study, certain important differences exist: It is, in this study by Waters et al., not the entirely new lesion as it was in the INTACT trial. Waters et al. define a minimal lesion as a stenosis of less than 20 % luminal narrowing. This is not a new lesion, it is a clear-cut existing stenosis although not hemodynamically effective. In these lesions then they find lesser progression per patient and per lesion. Unfortunately, the difference in distribution of smokers and the difference in age in the two groups may offset the conclusion given in this study. It at least makes the validation difficult. It is known that the cessation of smoking goes along with regression, whereas continuation to smoke allows progression of the lesions. On the other hand, at higher age progression is usually slower. The two factors therefore tend to confound the picture and not allow the conclusion that nicardipine halts progression or inhibits the appearance of new lesions.

The possible antiatheromatous effect of calcium channel blockers was tested in yet another model by Schroeder and coworkers (9, 10). They observed in a prospec-

Table 3. DB PC R CCAD QACG 13 months.

	Nifedipine	Placebo	
n =	168	167	
Progression	44 %	43 %	
Regression	23 %	21 %	
Minimal lesions:			
Progression-patient	15 %	37 %	*p* = 0.05
Progression-lesions	9 %	16 %	*p* = 0.05

Waters et al. Circulation, 1990
DB PC R: double-blind, placebo controlled, randomized design
CCAD: chronic coronary artery disease
QACG: quantitative angiocardiography

tive study the effects of diltiazem on the development of premature atheromatosis in the coronary system of transplanted hearts in humans. For comparison, the authors used heart-transplanted patients receiving the "normal" immunosuppressive treatment. They employed quantitative angiocardiography over a 2-year period with 240 – 270 mg of diltiazem/day in addition to the usual baseline treatment of heart-transplant patients with dipyridamol, acetylsalicylic acid, ciclosporine, azathioprin and prednisone in varying doses. 106 patients entered into the study, 54 received diltiazem, 52 served as controls after randomization. At 1 and 2 years, restudy was done; at 2 years, 33 patients still had quantitative angiographic assessment in the diltiazem group and 30 in the control group. In the comparison of the two groups at baseline, at 1 year and at 2 years, the average diameter for all coronary artery segments measured was with 2.41 mm in the control group and 2.32 mm in the diltiazem group, which was not statistically significantly different. At 1 year, however, a narrowing of the coronary arteries had occurred. This was more pronounced in the control group with 2.19 mm than in the diltiazem-treated group with 2.32 mm. The p-value reached 0.08 (Table 4). At 2 years, this difference was maintained and the vessels were about of the same size. The difference between the control and the diltiazem-treated group was small, with 0.14 mm; the p-value was 0.038. Comparing the patients with visible coronary arterial changes at 1 and at 2 years, there were none in the diltiazem group, and 10% in the control group. After 5 years, 20% of the diltiazem-treated patients showed visible coronary arterial changes indicative of atheromatosis, whereas in the control group this was about or more than 50%.

The authors conclude from their study that the calcium channel blocker diltiazem is able to delay the appearance of this peculiar type of atheromatosis in the coronary artery system of the transplanted heart under the current immunosuppressive treatment. The effect, as measurable from the size of the coronary arteries, however, is rather limited; in regard to the appearance of angiographically visible lesions, however, it is clearer.

From the studies published and discussed here, it can be concluded that, well in keeping with the experimental evidence, calcium channel blockers do seem to have a certain preventive effect on the arterial system in man, namely, on the coronary arterial system. Here, natural progression of the disease cannot be halted, regression cannot be induced. The appearance of new lesions can probably be retarded with clinically tolerated doses of nifedipine. Diltiazem may have a certain effect upon the coro-

Table 4. Diltiazem after HTX, R QACG 2 years (concomitant therapy: dipyridamol, ass, cyclosporine, azathioprin, prednisone).

	Diltiazem	Control
n =	54	52
Vessel diameter baseline	2.32	2.41 mm
1 year	2.32	2.19 mm
2 years	2.36	2.22 mm

Schroeder et al. Nejm, 1993
HTX: heart transplantation
R: randomized design
QACG: quantitative angiocardiography

nary arterial system of transplanted hearts in that it seems to retard the appearance of premature atheromatosis.

The preventive effect of nicardipine has not convincingly been demonstrated. Likewise, the studies published with verapamil do not seem to show such an effect. These studies have not been discussed here, because they have either not been published or the work published so far has been criticized in regard to the design and evaluation.

Finally, a brief mention shall be made of the effects of the calcium channel blockers on the mortality and event rates in chronic coronary artery disease:

Diltiazem: Gibson and coworkers (2) studied diltiazem in a double-blind, randomized, placebo-controlled, multicenter trial at a dose of 360 mg/day in patients with coronary artery disease with non-q wave infarction. They began 24 h after the infarct and followed the patients for 2 weeks. Reinfarction rate under placebo was 27 of 186 patients and in the diltiazem group 15 of 287 patients. Considering the very homogeneous contribution of the patients to the two groups, this effect was statistically highly significant. It can be concluded from the study that diltiazem does indeed reduce the infarction rate after non-q wave infarction during the first 2 weeks after the infarct.

In the large MDPIT trial (14), another double blind, randomized, placebo-controlled, multicenter trial, diltiazem was studied over a 2-year period in patients with chronic coronary artery disease after acute myocardial infarction. Nearly 2500 patients were studied over a 25-month period. Mortality was not different in the two groups. The cardiac event rate was, however, lower under diltiazem than under placebo. This relates to a reduced rate of reinfarction by 11 % in the diltiazem group. 202 patients of 1234 developed reinfarction in the verum group and 226 of 1232 did so under placebo.Therefore, here again a reduction of reinfarction is seen with diltiazem.

A further analysis of the cardiac event rate revealed that the presence or absence of pulmonary congestion influence the cardiac event rate remarkably: If pulmonary congestion was present, then the cardiac event rate was significantly higher under diltiazem than under placebo. Without pulmonary congestion cardiac event rate was lower with diltiazem than under placebo. It is to be concluded from this observation that the intrinsic negative inotropic effect of the calcium channel blocker may be disadvantageous in cases where there is evidence of cardiac failure.

Verapamil: The effect of verapamil on mortality and major cardiac events after an acute myocardial infarction was studied in a double blind, randomized, placebo-controlled, multicenter, Danish Verapamil Infarction Trial (DAVIT) (11, 12). 1775 patients were randomized. They were quite evenly distributed over the two groups. Verapamil was given in a dose of 360 mg/day and began in the 2nd week after the infarct and was continued for up to 18 months.

The cumulative mortality rate was slightly lower under verapamil, as was the major event rate. The difference does not reach with $p = 0.11$ or $p = 0.03$ statistical significance. Only after separation of the patients into a group with and without heart failure did statistically significant and meaningful differences emerged: Very much similar to what had been seen with diltiazem, the MDPIT study mortality rate and event rate were higher if heart failure was present. In the absence of heart failure verapamil seemed to improve cumulative mortality rate as well as event rate.

Nifedipine has been studied in numerous trials in acute myocardial infarction and in unstable angina pectoris. The experience has been disappointing (Muller et al., 1984). We have attempted to demonstrate a myocardial protective effect in acute

182

myocardial infarction (Just et al.). The result was statistically only marginal. The discrepancy to experimental evidence may be explained by the secondary sympathetic reflex activation. Simultaneous application of nifedipine with beta-blockers appears to be highly beneficial.

Myocardial protective effects of calcium antagonists have clearly been demonstrated for nifedipine in open-heart surgery (Just, Tschirkow, Schlosser 1982).

In conclusion, we deduce from the studies presented that the calcium channel blocker diltiazem is able to improve the clinical course after non-q-wave infarction by reduction of reinfarction rate. This effect is seen during the hospital course, as well as during the long term. In the presence of heart failure this effect disappears and the event rate increases under diltiazem. The observations for verapamil (1) are very similar in that the presence or absence of heart failure makes for a clear-cut difference: after acute myocardial infarction neither mortality nor event rate are influenced. If a separation is made for the presence or absence of heart failure, then a benificial effect is seen only if heart failure is absent.

In spite of the most convincing evidence for an antiatheromatous effect of calcium channel blockers in the clinical situation this effect can be attained only to a limited extent. Symptomatic relief may be most impressive. Mortality rates are, however, scarcely influenced (7, 8, 13, 14). The presence of heart failure precludes the application of calcium channel blockers. In the absence of heart failure a clinically important reduction of cardiac event rate may be seen.

A myocardial protective effect is seen in acute myocardial infarction and can be utilized in open-heart surgery.

References

1. Bussmann WD, Seher W, Gruengras M (1984) Reduction of creatine kinase and creatine kinase-MB indexes of infarct size by intravenous verapamil. Am J Cardiol 54: 1224–1230
2. Gibson RS, Boden WE, Theroux P, Strauss HD, Pratt CM, Gheorghiade M, Capone RJ, Crawford MH, Schlant RC, Kleiger RE et al. (1986) Diltiazem and reinfarction in patients with non-Q-wave myocardial infarction. Results of a double-blind, randomized, multicenter trial. N Engl J Med 315: 423–429
3. Gottlieb SO, Becker LC, Weiss JL, Shapiro EP, Chandra NC, Flaherty JT, Gottlieb SH, Ouyang P, Mellits ED, Townsend SN, Weisfeldt ML, Healy B, Gerstenblith G (1988) Nifedipine in acute myocardial infarction: an assessment of left ventricular function, infarct size, and infarct expansion. Br Heart J 59: 411–418
4. Gottlieb SO, Jeffrey A, Brinker E, Mellits D, Achuff SC, Baughman KL, Traill TA, Weiss JL, Reitz BA, Weisfeldt ML, Gerstenblith G (1989) Effect of nifedipine on the development of coronary bypass graft stenoses in high-risk patients: a randomized bouble-blind, placebo-controlled trial. Circulation 80: II–228
5. Lichtlen PR, Hugenholtz PG, Rafflenbeul W, Hecker H, Jost S, Deckers JW (1990) Retardation of angiographic progression of coronary artery disease by nifedipine. Results of the international nifedipine trial on antiatherosclerotic therapy (INTACT). INTACT Group Investigators. Lancet 335: 1109–1113
6. Loaldi A, Pollese A, Montorsi P, De Cesare N, Fabbiocchi F, Ravagnani P, Guazzi MD (1989) Comparison of nifedipine, propranolol and isosorbide dinitrate on angiographic progression and regression of coronary arterial narrowings in angina pectoris. Am J Cardiol 64: 433–439

7. Lubsen J, Tijssen JGP (1987) Efficacy of nifedipine and metoprolol in the early treatment of unstable angina in the coronary care unit: findings from the holland interuniversity nifedipine/metoprolol trial (HINT). Am J Cardiol 60: 18A–25A
8. Muller JE, Turi ZG, Pearle DL, Schneider JF, Serfas DH, Morrison J, Stone PH, Rude RE, Rosner B, Sobel BE, Tate C, Scheiner E, Roberts R, Hennekens CH, Braunwald E (1984) Nifedipine and conventional therapy for unstable angina pectoris: a randomized, double-blind comparison. Circulation 69: 728–739
9. John S, Schroeder MD, Shao-Zhou Gao MD, Sharon A, Hunt MD, Edward B, Stinson MD (1992) Accelerated graft coronary artery disease: diagnosis and prevention. J Heart Lung Transplant 11: 258–266
10. John S, Schroeder MD, Shao-Zhou Gao MD, Edwin L, Alderman MD, Sharon A, Hunt MD, Iain Johnstone PhD, Derek B, Boothroyd MSc, Voy Wiederhold MA, Edward B, Stinson MD (1993) A preliminary study of diltiazem in the prevention of coronary artery disease · in heart-transplant recipients. N Engl J Med 328: 164–170
11. The danish study group on verapamil in myocardial infarction (1986) The danish studies on verapamil in acute myocardial infarction. Br J Clin Pharmacol 21: 197S–204S
12. The danish study group on verapamil in myocardial infarction (1990) Effect of verapamil on mortality and major events after acute myocardial infarction (the danish verapamil infarction trial II – DAVIT II). Am J Cardiol 66: 779–785
13. The Israeli Sprint Study Group (1988) Secondary prevention reinfarction Israeli nifedipine trial (SPRINT). A randomized intervention trial of nifedipine in patients with acute myocardial infarction. Eur Heart J 9: 354–364
14. The multicenter diltiazem postinfarction trial research group (1988) The effect of diltiazem on mortality and reinfarction after myocardial infarction. N Engl J Med 319: 385–392
15. The SPRINT Study Group II (1988) The secondary prevention reinfarction Israeli nifedipine trial (SPRINT) II: Design and methods. Eur Heart J 9 (Suppl 1): 350 (Abstract)
16. Waters JD, Lespérance J, Francetich M, Causey D, Théroux P, Chiang YK, Hudon G, Lemarbre L, Reitman M, Joyal M, Gosselin G (1990) A controlled clinical trial to assess the effect of a calcium channel blocker on the progression of coronary atherosclerosis. Circulation 82: 1940–1953
17. Just H, Heidelbach I, Jäckle B, Wollschläger H (1986) Nifedipin bei akutem Myokardinfarkt: Ist ein myokard-protektiver Effekt in einer doppelt-blinden, placebokontrollierten Studie nachweisbar? Intensivmedizin 23: 159
18. Just H, Tschirkow A, Schlosser V (1982) Kalziumantagonisten zur Kardioplegie und Myokardprotektion in der offenen Herzchirurgie. G. Thieme Verlag Stuttgart New York

Author's address:
Prof. Dr. H. Just
Medizinische Universitätsklinik
Abt. Innere Medizin III
Kardiologie und Angiologie
Hugstetter Straße 55
D-79106 Freiburg
FRG

Lipid-lowering therapy –
Implications for the prevention of atherosclerosis

G. Schmitz, K. J. Lackner

Institut für Klinische Chemie und Laboratoriumsmedizin, Universität Regensburg, FRG

Summary: Lipid-lowering therapy has been advocated as a means to decrease the incidence of cardiovascular disease. The basis for this are results of animal studies, epidemiologic surveys, and intervention studies in man. There have been recommendations by national and international advisory boards regarding the aims and means of such interventional strategies. These have not been unanimously accepted, however. The present article will review the evidence on which intervention strategies may be based and the current status of the scientific discussion. The implications for future strategies to prevent atherosclerosis are discussed.

Key words: Lipoproteins – prevention of atherosclerosis – lipid-lowering therapy

Introduction

Lipid-lowering therapy usually has either of two goals. The major goal is the prevention of cardiovascular disease (CVD) or its progression. There is broad consensus that increased cholesterol levels and specific other hyper- and dyslipidemias are related to an increased cardiovascular risk (e.g., 12,60). However, there is no general consensus on the therapeutic implications. There have been numerous recommendations of national and international committees or task forces for lipid-lowering treatment strategies to reduce cardiovascular morbidity (13, 21, 55, 63–65). None of these recommendations are undisputed (32, 44, 48). In another patient group, prevention of pancreatitis, which may be caused by highly elevated triglycerides, is the reason for lipid-lowering intervention. It is widely accepted that triglycerides above 600 mg/dl are related to a substantially increased risk for acute pancreatitis. The aim of this paper is to review the current evidence for lipid-lowering therapy to prevent atherosclerosis and discuss future prospects of diagnosis and therapy of hyperlipidemia.

Epidemiology and intervention

Numerous epidemiologic surveys have shown a positive correlation between the level of plasma cholesterol and the risk for CVD (e.g. 12,60). In addition, it has been shown that low-density lipoprotein (LDL)-cholesterol is positively correlated to CVD, whereas the level of high-density lipoprotein (HDL)-cholesterol is inversely correlated to cardiovascular risk (26, 27). A specific modified apoB-containing particle, lipoprotein Lp(a) has also been identified as an independent risk factor (18, 68).

In addition, it appears that certain forms of hypertriglyceridemia, namely, those associated with low HDL-cholesterol have a substantially increased risk for CVD. And finally, there are several genetic disorders of lipoprotein metabolism with an increased risk for CVD. These include familial hypercholesterolemia (FH) (24), familial combined hyperlipoproteinemia (9, 25), familial dysbetalipoproteinemia (Type III hyperlipoproteinemia) (42), familial dyslipidemic hypertension (69), and several very rare monogenic disorders (Table 1). There have been claims that very low plasma cholesterol levels are associated with an increased mortality, particularly from cancer. This issue has been reviewed recently (34). In several but not all surveys the association of low cholesterol and cancer disappeared, if cancer cases diagnosed within 2 years from entry into the study were excluded. Since it is known that malignancies may lower cholesterol, this was interpreted to indicate that the increased incidence of cancer in the cohort with the lowest cholesterol levels was related to preexisting disease at the time of lipid screening.

Given the epidemiologic evidence, the question remained whether modifying a supposedly atherogenic lipoprotein profile could reduce the risk for CVD. Several trials addressing this question have been conducted up to now. They may be grouped into three major categories. 1) Primary prevention trials in probands without any evidence for preexisting CVD, 2) secondary prevention trials in patients with known CVD, and 3) angiographically controlled trials in patients with CVD.

The primary prevention trials have several aspects in common (Table 2). All have been conducted in men between 40 and 65 years of age with increased cholesterol values or increased apoB-containing particles (7, 22, 37, 38, 49). Cardiovascular morbidity was significantly reduced in the drug trials. The major ciriticism has been that

Table 1. Familial Hyperlipoproteinemias.

Hyperlipo-proteinemia	Genetic defect	Lipoprotein pattern	Cardiovascular risk
Familial hypercholesterolemia	LDL-receptor	LDL-chol ↑ ↑	↑ ↑ ↑ ↑
Familial defective apolipoprotein B	ApoB-100 $Gln_{3500} \rightarrow Arg$	LDL-chol ↑	↑ ↑ ↑
Familial combined hyperlipoproteinemia	?	LDL ↑ and/or VLDL ↑	↑ ↑ ↑
Familial dyslipidemic hypertension	?	LDL ↑ and/or VLDL ↑	↑ ↑
Type III hyperlipoproteinemia	ApoE 2/2	β-VLDL ↑	↑ ↑ ↑ ↑
Familial chylo-micronemia	LPL ApoC-II	chylomicrons triglycerides ↑ ↑ ↑	pancreatitis
Hypoalphali-poproteinemia	heterogeneous some known	HDL-Chol ↓	→ to ↑ ↑ ↑ depending on defect
Hypoalphalipoproteinemia with hypertriglyceridemia	?	HDL-chol ↓ and VLDL ↑	↑ ↑

Table 2. Cause-specific mortality in primary and secondary prevention trials.

Trial	n		All-causes		Coronary heart disease		Cancer		Violence/ Accidents	
	C	T	C	T	C	T	C	T	C	T
Primary Prevention										
LRC-CPPT	1.900	1.906	3.7	3.6	2.0	1.6	0.8	0.8	0.2	0.6
Helsinki	2.030	2.051	2.1	2.2	0.9	0.7	0.5	0.5	0.2	0.5
WHO-Coop	5.296	5.331	1.6	2.4	0.64	0.68	0.5	0.8	0.28	0.33
Secondary Prevention										
POSCH	417	421	14.9	11.6	10.6	7.6	1.9	1.9	0.7	0.7
CDP	2.789	1.119	58.2	52.0*	41.3	35.5*	4.4	4.0	1.1	0.9
Stockholm	276	279	29.7	21.9*	26.4	16.8*	2.2	1.4	n.a.	n.a.

C: controls; T: treated; *: statistically significant difference

none of these trials could show a survival benefit for the treatment group. In this respect, several authors have pointed to the fact that in the three major primary prevention trials the reduction in cardiovascular mortality was always offset by excess deaths from other causes (14, 33, 46). This has been interpreted as indicating that the excess mortality is in some way related to the intervention itself. This would have major implications for the extrapolation of the data to a population at lower risk (i.e., women, persons with lesser degrees of hyperlipoproteinemia), because one might argue that the overall outcome in such a group might in fact be detrimental. One other point of criticism is the high cost of drug intervention. The effects in terms of life expectancy, even in a group which is supposedly at high risk, were at best marginal. Intervention in a group at lower risk would be even more expensive, with a smaller impact on life expectancy (61, 66).

An alternative to drug intervention is dietary intervention. However, the results of primary prevention trials with diet alone have been somewhat disquietening, because in some but not all trials there was an increased cancer incidence (1, 7, 20, 34, 35, 67). The reason for this observation is unclear. It cannot be ruled out that very rigorous lipid-lowering diets may have undesired side-effects. This may be particularly relevant for diets rich in polyunsaturated fatty acids. Furthermore, the results of dietary intervention are dependent on the motivation and compliance of the population studied. In a "freeliving" population with little guidance the results in terms of lipid-lowering have usually been disappointing (52). Thus, dietary intervention appears to be less important for high-risk strategies, but may be more useful for population-based approaches, i.e., attempts to slowly modify the diet of everyone irrespective of any preexisting risk for CVD.

Thus, it becomes obvious that one goal in *primary* prevention has to be an improved diagnostic approach to better define the patients at increased risk which will benefit most from intervention (29).

The majority of patients in the secondary prevention trials were also men. However, the number of women included is large enough to draw conclusions. It appears that there are no relevant differences in terms of risk and risk reduction between men and women in these trials. In two of these trials a significant reduction of overall mortality was achieved (8, 10, 11, 15). This is related to the fact that CVD is the major cause of death in this preselected group. In addition, mortality in these trials was much higher than in the primary prevention trials giving the trials a larger statistic power. There is no doubt that lipid lowering therapy in patients with preexisting CVD reduces cardiovascular morbidity and overall mortality. The data even indicate that the lipid level before therapy need not be elevated to obtain a risk reduction (10).

In the angiographic trials a reduction in the progression of CVD could be shown with lipid lowering. Furthermore, there appeared to be more regression in the intervention groups compared to the controls. Both effects appear to be related to the degree of lipid lowering rather than to the means of lipid-lowering (3, 5, 36).

Thus, the prevention trials indicate that the presence of CVD is currently one of the most efficient indicators to define a subpopulation of the general population, which will benefit most from lipid-lowering therapy. Primary prevention with drugs based on lipid levels alone is an expensive undertaking. In regard of the tremendous increases in health care expenses in all industrialized countries, cost-effectiveness is an important issue at a time of limited resources. Extrapolation to populations at lower risk than the probands in the intervention studies is not feasible since there are no data available regarding excess risk conveyed by the therapeutic intervention. The impact of defining apparently healthy individuals as patients who require treatment has only recently been considered (32). In fact, analysis in hypertension indicates that this leads to increased absenteeism and decreased psychosocial function (30, 45). Assessments of quality-of-life changes brought about by lipid-lowering interventions have not been performed. Especially, it is unknown whether new treatment strategies and drugs interfere less with quality of life than previous interventions, as one might expect. And finally, the books on dietary intervention are not yet closed.

Improvement of diagnostic strategies

It is obvious that every adult should have his or her cholesterol and triglyceride level checked at least once. Whether HDL-cholesterol should be determined at the initial screening visit is an unanswered question in terms of cost effectiveness. Any further determination cannot be recommended for screening purposes, but should be restricted to specific patients. Children should be analyzed only when there is a family history of hyper- or dyslipoproteinemia or when there are physical signs of a disorder of lipid metabolism (e.g., xanthomas, arcus corneae) (39, 47).

To better define patients at an increased cardiovascular risk, more sophisticated diagnostic approaches are needed. Moreover, there is a need for additional diagnostic markers which allow therapeutic monitoring of antiatherogenic treatment.

First, one has to consider the cellular mechanisms of atherogenesis (56). The cells involved in this process are monocytes, T-lymphocytes, platelets, endothelial cells, and smooth muscle cells (SMC). The typical early atherosclerotic lesion consists of an accumulation of monocyte-derived macrophages in the intima which store large amounts of lipid. In addition, T-lymphocytes are present in the lesion. When lesions advance there is usually an increased number of intimal SMCs. This is accompanied by an increase in fibrotic material. This lesion is called a fibromuscular plaque. How-

188

ever, in many advanced lesions there are typical fibrotic regions adjacent to lipid-rich areas. One important functional difference between these two morphologically different lesions may be the higher tendency for plaque rupture of the soft lipid-rich plaque. The fibrous lesion is usually very stable and resistant to shear stress. Another feature of atherosclerosis in the medium sized muscular artery is neovascularization of the vessel wall. Whether the cells of the newly grown vessels take active part in the development of the plaque is a matter of intensive investigation.

One diagnostic approach for the detection of patients with progressing atherosclerosis is to define monocyte and possibly T-lymphocyte markers or subpopulations in the blood which change in the process of intimal infiltration, i.e. during lesion development. The identification of such markers or subpopulations would permit to diagnose patients with active lesion development.

Data from our group show that hypercholesterolemia leads to a changed pattern of surface marker expression in monocytes compared to controls depending on the cause of hypercholesterolemia. It could be shown that monocytes from patients with homozygous FH had a lower expression of the differentiation markers CD 14, CD 16, and HLA-DR than controls. A similar observation was made in heterozygous FH patients. However, the degree of reduced expression was less. In contrast, non-FH hypercholesterolemic patients have an increased expression of these markers. In other studies, it was possible to differentiate between hypercholesterolemic patients with and without cardiovascular disease on the basis of monocyte surface marker expression. In hypercholesterolemic patients with cardiovascular disease a specific CD 14^+/CD 16^+ monocyte subpopulation was reduced, indicating a more rapid disappearance of these cells into a non-vascular pool, possibly the arterial wall. If these results hold true in a larger collective, this approach may permit to define patients at an increased risk for cardiovascular disease or with actively progressing vascular disease.

It is hypothesized that oxidative modification of apoB containing lipoproteins renders them much more atherogenic (50, 62). These modified lipoproteins have several biological activities different from native lipoproteins. In particular, minimally oxidized LDL can induce the expression of adhesion molecules in endothelial cells as well as the production of chemotactic factors and cytokines in monocytes, endothelial cells and smooth muscle cells (2, 17, 51). If LDL particles are further oxidized, they become cytotoxic. In addition, they are taken up by macrophages in an uncontrolled manner leading to foam cell formation (62). Thus, the analysis of the status of redox systems in the blood and in blood cells may prove useful in the determination of cardiovascular risk. This may include determination of thiobarbituric acid reactive substances (TBARS) in lipoproteins, fluorescence of lipoproteins, detection of oxidized lipids, but also the analysis of redox systems like vitamins E and C, oxidating substances like homocysteine and others. Another modification of lipoproteins which might render them more atherogenic is nonenzymic glycation (57). The end product are AGE (advanced glycosylation end products)-lipoproteins. Thus, determination of glycated lipoproteins may prove useful in the determination of cardiovascular risk in diabetics.

Since several familial dyslipoproteinemias carry a high risk for CVD it will be important to effectively diagnose these patients. This includes determination of the LDL-receptor, which can be performed by a simple flow cytometric method developed in our laboratory. Perhaps in the future, it will be possible to determine other lipoprotein receptors like the LDL-receptor related protein (LRP), the scavenger receptors, and HDL-receptors.

Determination of Lp(a) as an independent risk factor is important. Plasma levels can be routinely determined. In addition, the analysis of the apo(a) genotype which may influence the atherogenic potential of the Lp(a) particle can be performed.

Mutants of different apolipoproteins, enzymes involved in lipoprotein metabolism (e.g. lipoprotein lipase, lecithin: cholesteryl acyltransferase) and transfer proteins (cholesteryl ester transfer protein) can be analyzed already. Some of these mutations are so rare that they are only relevant for specialized centers. Examples for this type of mutation are deficiencies of apolipoprotein A-I, CETP-deficiency, and LCAT-deficiency. There are only a few patients worldwide with these disorders. However, some defects are fairly common as, for instance, a mutation in the apoB gene substituting a glutamine for an arginine at position 3500 that leads to defective binding of the mutant protein, apoB-3500, to the LDL-receptor and to hypercholesterolemia (59). This defect has been estimated to have a heterozygote frequency of approx. 1/500 in the general population (53). Another important example is several fairly common mutations at the LPL locus that affect triglyceride levels, postprandial lipemia, and potentially the risk for CVD.

ApoE is a polymorphic plasma protein with three major alleles: ε2, ε3, and ε4 (41). The ε3 allele is the most common with approx. 80 % (19). In the population there are three homozygous and three heterozygous apoE-phenotypes. The apoE-penotype has a strong influence on total and LDL-cholesterol with apoE2/2 homozygotes having the lowest LDL-cholesterol and apoE4/4 homozygotes having the highest LDL-cholesterol (19). In addition, the presence of apoE2 (also in combination with apoE3 or apoE4) appears to make individuals more sensitive to diet-induced hypertriglyceridemia, wheres apoE4 renders them more sensitive to diet-induced hypercholesterolemia (4, 31, 40). Thus, apoE-phenotyping has a direct impact on dietary counseling of patients. The diagnosis of familial dysbetalipoproteinemia depends on the analysis of apoE isoforms, because this disorder is almost exclusively seen in homozygous individuals for apoE2. These constitute approximately 1 % of the population. However, only about one in 50 apoE2 homozygotes develops familial dysbetalipoproteinemia. It appears that a second, yet unknown genetic factor or possibly specific environmental factors are required for the disease (19). E2 homozygotes usually have atherogenic, cholesterol enriched β-VLDL particles in fasting plasma, regardless of whether they have type III hyperlipoproteinemia or not. There is evidence that some E2/3 and E2/4 heterozygotes, which constitute approximately 12–15 % of the population, have also an increased prevalence of β-VLDL. On the other hand, E2/3 heterozygosity appears to protect from cardiovascular disease (16, 43).

Lipid-lowering therapy

Which patients should obtain lipid-lowering interventions? As a guideline, the latest recommendations of the European Atherosclerosis Society may be used (21). These may be subdivided in the "Management of hypercholesterolemia", "Management of mixed hyperlipidemia" and "Management of hypertriglyceridemia". The recommendations have been modified, particularly with respect to the treatment goals (Table 3). The very low LDL-cholesterol levels recommended in earlier publications of this society have not been kept for every patient (63, 64). However, the cholesterol target levels for men are 195 mg/dl for total cholesterol and 135–155 mg/dl for LDL-choles-

Table 3. Recommendations of the European Atherosclerosis Society (modified from [21]).

Global risk	Target level	
	Plasma chol.	**LDL-chol.**

Mildly increased
    pretreatment cholesterol 200–300 mg/dl
    no non-lipid risk factors
    Plasma Cholesterol/HDL-cholesterol 4.5–5.0
	195–230 mg/dl	155–175 mg/dl

Moderately increased
    pretreatment cholesterol 200–300 mg/dl
        PLUS one non-lipid risk factor or
        PLUS HDL-cholesterol < 30 mg/dl
	195 mg/dl	135–155 mg/dl

Highly increased
    presence of vascular disease
    presence of FH or plasma chol. < 300 mg/dl
    pretreatment cholesterol 200–300 mg/dl
        PLUS two non-lipid risk fctors or
        PLUS one severe non-lipid risk factor
	175–195 mg/dl	115–135 mg/dl

terol assuming that male sex constitutes a non-lipid risk factor. This means that more than 50 % of the male population would need some form of therapy. For women the target levels are 230 mg/dl and 175 mg/dl for total cholesterol and LDL-cholesterol, respectively.

One should keep in mind that these recommendations are based on extrapolations from epidemiologic surveys and intervention trials. They imply that the recommended intervention has no undesired effects. These could become relevant if the effect on cardiovascular morbidity is only small, as one might expect from a cholesterol lowering from, for example, 220 mg/dl to 190 mg/dl.

One can assume that in an unselected population with dietary measures at best a 10–20 % reduction of plasma cholesterol can be achieved (6, 28, 54). This implies that most men with a total cholesterol above 240 mg/dl and many above 220 mg/dl will need drug treatment to reach the target levels. To achieve a more pronounced lipid lowering effect one has to institute very rigorous dietary regimens which will not be followed by the majority of patients. Even if they were, there remain some unanswered questions: 1) does dietary (or drug induced) lipid-lowering reduce cardiovascular and overall mortality in otherwise healthy individuals with total cholesterol below 250 mg/dl?; 2) are lipid-lowering diets with a high proportion of polyunsaturated fatty acids safe? Both questions are still debated. Arguments against this approach point to the fact that there has been one large trial reported in which dietary lipid-lowering was linked to an increased cancer incidence. Polyunsaturated fatty acids have been shown to promote carcinogenesis in animal models (34). This has led to the recommendation by some authorities to substitute saturated fatty acids by the monounsaturated oleic acid. This is equally effective in lipid-lowering and there is no evidence of untoward effects in large populations (Mediterranean countries) (28).

There are several points of criticism against a population-based intervention strategy. It is obvious that an intervention strategy aimed at high-risk patients will be much more efficient. Furthermore, it will be unlikely that the benefits of lipid-lowering are offset by its side effects in a group at high risk for CVD. We would therefore argue that until more data are available, drug treatment for primary prevention should be restricted to those groups that have been included in the intervention trials. That means, men with a total cholesterol above 260 mg/dl or an LDL-cholesterol above 190 mg/dl. Reasonable exceptions from this rule may be patients with at least two additional non-lipid risk factors and patients with familial disorders of lipid metabolism. The treatment of women with moderately elevated cholesterol has still to be shown to be beneficial and cost effective. People above the age of 70 without any evidence of cardiovascular disease should not be screened and treated only rarely, because there have been calculations showing that a screening and treatment program for such a population will be very costly with little if any effect on morbidity and life expectancy (23). In fact, there is evidence that FH patients reaching the 7th decade of life have the same cardivascular risk as age-matched controls. This argues for a natural resistance against cholesterol induced atherosclerosis in these FH-patients (58). The same might be true for other elderly hypercholesterolemic patients, because LDL-cholesterol is not well correlated to cardiovascular risk in old people. There is no evidence available that lipid-lowering therapy of mild hyperlipidemia in children has any beneficial effect. Whether major dietary changes are safe or will cause developmental problems has not been assessed in larger trials. The treatment of children should, therefore, be restricted to familial hypercholesterolemia and other rare genetic disorders. These children should be followed in specialized centers.

Dietary counseling should emphasize the reduction of total fat and cholesterol intake. The recommendation of a high intake of polyunsaturated fatty acids should not be made, however, the substitution of saturated fatty acids by monounsaturated fatty acids may be recommended since there are large populations in the Mediterranean countries on such a diet without evidence for harmful side-effects of these oils.

In patients with known CVD intervention should be much more aggressive. Since the available data indicate that these patients benefit independent from their pretreatment lipoprotein levels, one should aim at very strict target levels. The treatment goal should be an LDL-cholesterol below 135 mg/dl and normalization of triglycerides if they are increased. To this end dietary and drug intervention should be used. If necessary, these patients may be treated with drug combinations.

Clearly, all patients with triglycerides above 600 mg/dl should be treated to prevent acute pancreatitis. As stated by the European Atherosclerosis Society, hypertriglyceridemia usually responds well to dietary measures and weight reduction.

Having defined the patients to be treated, the mode of therapy depends mainly on the lipoprotein pattern of the patient. Every patient considered for therapeutic intervention should have dietary counseling before any other intervention. Except for hyperchylomicronemia (Type I hyperlipoproteinemia) and other forms of severe hypertriglyceridemia dietary intervention is rather uniform. The aim is to reduce total fat and cholesterol intake and substitute this with complex carbohydrates. The amount of dietary fiber should be increased. The percentage of total caloric intake as fat should not exceed 30 %. Cholesterol intake should be restricted to 300 mg/day or less. The percentage of unsaturated fatty acids should be increased. However, it appears not to be prudent to increase the intake of polyunsaturated fatty acids to as high as 1/3 of total fatty acids. An alternative is to increase the relative amount of oleic acid (olive oil) in the diet (4, 24, 46). Some patients will need an initial weight reduction. One should be aware that dietary compliance is inversely correlated to palatability and therefore to dietary fat content. If dietary measures do not lead to a satisfactory reduction in lipids, drugs may be considered (Table 4).

For hypercholesterolemia with elevated LDL-cholesterol HMGCoA-reductase inhibitors and ion exchange resins are the drugs of choice. Depending on the severity of hypercholesterolemia and the targeted cholesterol value these drugs may be given alone or in combination. It should be kept in mind that women of childbearing age should not receive HMGCoA-reductase inhibitors unless adequate contraceptive measures have been taken. In addition, one should be cautious about the use of HMGCoA-reductase inhibitors in children or young adults as long as long-term safety data are not available. Today, there is no compelling evidence that one or another reductase inhibitor is advantageous as long as equipotent doses are used.

If elevated LDL-cholesterol is accompanied by hypertriglyceridemia fibric acid derivatives, nicotinic acid or reductase inhibitors may be used. However, the combination of fibrates or nicotinic acid with HMGCoA-reductase inhibitors should be restricted to patients who cannot be treated otherwise and should be supervised in specialized centres. This recommendation is based on the fact that there is a high incidence of myopathy with fibrates and HMGCoA-reductase inhibitors. The combination of nicotinic acid and HMGCoA-reductase inhibitors may be more hepatotoxic than the single drugs. The addition of an ion exchange resin may prove useful to further lower LDL-cholesterol. However, this will usually increase triglycerides.

Hypertriglyceridemia alone only rarely needs drug treatment. If dietary compliance is good, this condition can be controlled without medication. If medication is considered, fibrates and nicotinic acid are the agents of choice. Sometimes these patients respond well to fish-oil substitution. However, one should be aware that the

Table 4. Relative potency of lipid-lowering agents in the treatment of different forms of hyperlipidemia.

	Hyper-cholesterolemia	Mixed hyperlipidemia	Hyper-triglyceridemia	Comment
HMGCoA-Reductase Inhibitors	+++	++	O	
Ion Exchange Resins	++	(+)	--	Increase triglycerides
Fibrates	+	++	++	
Nicotinic Acid	++	++	++	
β-Sitosterol	+	+	O	
Neomycin	++	+	O	Lowers Lp(a) Relevant toxicity
Probucol	+	O	O	Antioxidant efficacy perhaps more relevant

substitution of fish oil may substantially add to the daily caloric intake. Thus, it is important to really substitute other fats with fish oil.

All other lipoprotein disturbances either cannot be treated with drugs at this time or there are no data that treatment changes the risk for CVD. One example is low HDL-cholesterol. Today, there is no evidence available in man that increasing low HDL-cholesterol levels reduces cardiovascular risk. Lp(a) may be reduced by treatment with neomycin and nicotinic acid. However, no long-term data on the toxicity and side-effect of such a regimen are available. In addition, it is not known if lowering of Lp(a) will decrease cardiovascular risk.

Very few patients should be treated with lipoprotein apheresis. These include patients homozygous for FH. Heterozygous FH patients may be considered for treatment, if they do not respond adequately to drug treatment and have evidence of cardiovascular disease. Specific apheresis of Lp(a) may be considered in patients with isolated elevation of Lp(a), CVD and no other risk factors for CVD. This treatment is still experimental.

Future directions

One interesting therapeutic approach for the future will be to make apoB-particles more resistant to oxidative modification or to reduce prooxidant acitivity in general. This should reduce their atherogenicity without the necessity to lower plasma cholesterol. It remains to be shown that this approach, which appears to be successful in LDL-receptor deficient rabbits is also useful in man. A decrease in the hepatic production of apoB-containing particles may be another approach.

Independent from modifications of lipoprotein metabolism targets of anti-atherosclerotic therapy will be the cells involved in atherogenesis. This therapy may be monitored by invasive procedures such as angiography or intravascular ultrasound. Perhaps it will become feasible to monitor this therapy with non-invasive imaging procedures. One major aim for the future, however, will be to develop laboratory tests to assess the activity of CVD. Flow cytometry provides the technical means to assess the activation status of white blood cells. This approach may eventually permit to identify patient subgroups with active CVD.

References

1. Arntzenius AC, Kromhout D, Barth JD, Reiber JHC, Bruschke AVG, Buis B, van Gent CM, Kempen-Voogd N, Strikwerda S, van der Velde EA (1985) Diet, lipoproteins, and the progression of coronary atherosclerosis. The Leiden Intervention Trial. N Engl J Med 312: 805–811
2. Berliner JA, Territo MC, Sevanian A, Ramin S, Kim JA, Bamshad B, Esterson M, Fogelman AM (1990) Minimally modified low density lipoprotein stimulates monocyte endotelial interactions. J Clin Invest 85: 1260–1266
3. Blankenhorn DH, Nessim SA, Johnson RL, Sanmarco ME, Azen SP, Cashin-Hemphill L (1987) Beneficial effects of combined colestipol-niacin therapy on coronary atherosclerosis and coronary venous bypass grafts. JAMA 257: 3233–3240

4. Brenninkmeijer BJ, Stuyt PH, Demacker PN, Stalenhoef AF, van't Laar A (1987) Catabolism of chylomicron remnants in normolipidemic subjects in relation to the apoprotein E phenotype. J Lipid Res 28: 361–370
5. Brown G, Albers JJ, Fisher LD, Schaefer SM, Lin JT, Kaplan C, Zhao XQ, Bisson BD, Fitzpatrick VF, Dodge HT (1990) Regression of coronary artery disease as a result of intensive lipid-lowering therapy in men with high levels of apolipoprotein B. N Engl J Med 323: 1289–98
6. Brown WV (1990) Dietary recommendations to prevent coronary heart disease. Ann N Y Acad Sci 598: 376–388
7. Buchwald H, Fitch L, Moore RB (1982) Overview of randomized clinical trials of lipid intervention for atherosclerotic cardiovascular disease. Contr Clin Trials 3: 271–283
8. Buchwald H, Varco RL, Matts JP, Long JM, Fitch LL, Campbell GS, Pearce MB, Yellin AE, Edmiston WA, Smink RD Jr, Sawin HS Jr, Campos CT, Hansen BJ, Tuna N, Karnegis JN, Sanmarco ME, Amplatz K, Castaneda-Zuniga WR, Hunter DW, Bissett JK, Weber FJ, Stevenson JW, Leon AS, Chalmers TC, and the POSCH group (1990) Effect of partial ileal bypass surgery on mortality and morbidity from coronary heart disease in patients with hypercholesterolemia. Report of the Program on the Surgical Control of the Hyperlipidemias (POSCH). N Engl J Med 323: 946–955
9. Brunzell JD, Schrott HG, Motulsky AG, Bierman EL (1976) Myocardial infarction in the familial forms of hypertriglyceridemia. Metabolism 25: 313–320
10. Canner PL, Berge KG, Wenger NK, Stamler J, Friedman L, Prineas RJ, Friedewald W (1986) Fifteen-year mortality in coronary drug project patients: long-term benefit with niacin. J Am Coll Cardiol 8: 1245–1255
11. Carlson LA, Rosenhamer G (1988) Reduction of mortality in the Stockholm Ischaemic Heart Disease Secondary Prevention Study by combined treatment with clofibrate and nicotinic acid. Acta Med Scand 223: 405–424
12. Castelli WP, Garrison RJ, Wilson PWF, Abbott RD, Kalousdian S, Kannel WB (1986) Incidence of coronary heart disease and lipoprotein cholesterol levels. The Framingham Study. JAMA 256: 2835–2838
13. Consensus Conference (1985) Lowering blood cholesterol to prevent heart disease. JAMA 253: 2080–2086
14. Criqui MH (1991) Cholesterol, primary and secondary prevention, and all-cause mortality. Ann Intern Med 119: 973–976
15. The Coronary Drug Project Research Group (1975) Clofibrate and niacin in coronary heart disease. JAMA 231: 360–381
16. Cumming AM, Robertson F (1984) Polymorphism of the apo E locus in relation to risk of coronary disease. Clin Genet 25: 310–313
17. Cushing SD, Berliner JA, Valente AJ, Territo MC, Navab M, Parhami F, Gerrity R, Schwartz CJ, Fogelman AM (1990) Minimally modified low density lipoprotein induces monocyte chemotactic protein 1 in human endothelial cells and smooth muscle cells. Proc Natl Acad Sci USA 87: 5134–5138
18. Dahlen GH, Guyton JR, Attar M, Farmer JA, Kautz JA, Gotto AM Jr (1986) Association of levels of lipoprotein Lp(a), plasma lipids, and other lipoproteins with coronary artery disease documented by angiography. Circulation 74: 758–765
19. Davignon J, Gregg RE, Sing CF (1988) Apolipoprotein E polymorphism and atherosclerosis. Arteriosclerosis 8: 1–21
20. Ederer F, Leren P, Turpeinen O, Frantz ID Jr (1971) Cancer among men on cholesterol-lowering diets. Experience from five clinical trials. Lancet II: 203–206
21. European Atherosclerosis Society (1992) Prevention of coronary heart disease. Scientific background and new clinical guidelines. Nutr Metab Cardiovasc Dis 2: 113–154
22. Frick MH, Elo O, Haapa K, Heinonen OP, Heinsalmi P, Helo P, Huttunen JK, Kaitaniemi P, Koskinen P, Manninen V, Mäenpäa H, Mälkönen M, Mänttäri M, Norola S, Pasternack A, Pikkarainen J, Romo M, Sjöblom T, Nikkilä EA (1987) Helsinki heart study: primary-prevention trial with gemfibrozil in middle-aged men with dyslipidemia. N Engl J Med 317: 1237–1245

23. Garber AM, Littenberg B, Sox HC Jr, Wagner JL, Gluck M (1991) Costs and health consequences of cholesterol screening for asymptomatic older Americans. Arch Intern Med 151: 1089–1095

24. Goldstein JL, Brown MS (1989) Familial hypercholesterolemia. In: The metabolic basis of inherited disease. Scriver CR, Beaudet AL, Sly WS, Valle D (eds), McGraw-Hill, pp 1215–1250

25. Goldstein JL, Schrott HG, Hazzard WR, Bierman EL, Motulsky AG (1973) Hyperlipidemia in coronary heart disease. II. Genetic analysis in 176 families and delineation of a new inherited disorder, combined hyperlipidemia. J Clin Invest 52: 1544–1568

26. Gordon DJ, Probstfield JL, Garrison RJ, Neaton JD, Castelli WP, Knoke JD, Jacobs DR, Bangdiwala S, Tyroler HA (1989) High-density lipoprotein cholesterol and cardiovascular disease. Circulation 79: 8–15

27. Gordon T, Castelli WP, Hjortland MC, Kannel WB, Dawber TR (1977) High density lipoprotein as a protective factor against coronary heart disease. Am J Med 62: 707–714

28. Grundy SM (1987) Dietary therapy for different forms of hyperlipoproteinemia. Circulation 76: 523–528

29. Grundy SM (1990) Cholesterol and coronary heart disease. Future directions. JAMA 264: 3053–3059

30. Haynes RB, Sackett DL, Taylor DW, Gibson ES, Johnson AL (1978) Increased absenteeism from work after detection and labelling of hypertensive patients. N Engl Med 299: 741–744

31. Kesäniemi YA, Ehnholm C, Miettinen TA (1987) Intestinal cholesterol absorption efficiency in man is related to apolipoprotein E phenotype. J Clin Invest 80: 578–581

32. Krahn M, Naylor CD, Basinski AS, Detsky AS (1991) Comparison of an aggressive (U.S.) and a less aggressive (Canadian) policy for cholesterol screening and treatment. Ann Int Med 115: 248–255

33. Kronmal RA (1985) Commentary on the published results of the Lipid Research Clinics Primary Prevention Trial. JAMA 253: 2091–2093

34. Lackner KJ, Schettler G, Kübler W (1989) Plasma cholesterol, lipid lowering, and risk for cancer. An update of the results from epidemiologic studies and intervention trials. Klin Wochenschr 67: 957–962

35. Leren P (1970) The Oslo Diet-Heart Study. Circulation 42: 935–942

36. Levy RI, Brensike JF, Epstein SE, Kelsey SF, Passamani ER, Richardson JM, Loh IK, Stone NJ, Aldrich RF, Battaglini JW, Moriarity DJ, Fisher ML, Friedman L, Friedewald W, Detre KM (1984) The influence of changes in lipid values induced by cholestyramine and diet on progression of coronary artery disease: the results of the NHLBI type II coronary intervention study. Circulation 69: 325–337

37. The Lipid Research Clinics Program (1984) The lipid research clinics coronary primary prevention trial results. I. Reduction in incidence of coronary heart disease. JAMA 251: 351–364

38. The Lipid Research Clinics Program (1984) The lipid research clinics coronary primary prevention trial results. II. The relationship of reduction in incidence of coronary heart disese to cholesterol lowering. JAMA 251: 365–374

39. Lloyd JK (1991) Cholesterol: should we screen all children or change the diet of all children. Acta Paediatr Scand suppl 373: 66–72

40. Lussier-Cacan S, Bouthillier D, Davignon J (1985) ApoE allele frequency in primary endogenous hypertriglyceridemia (type IV) with and without hyperapobetalipoproteinemia. Arteriosclerosis 5: 639–643

41. Mahley RW (1988) Apolipoprotein E: cholesterol transport protein with expanding role in cell biology. Science 240: 622–630

42. Mahley RW, Rall SC Jr (1989) Type III hyperlipoproteinemia (dysbetalipoproteinemia): the role of apolipoprotein E in normal and abnormal lipoprotein metabolism. In: The metabolic basis of inherited disease. Scriver CR, Beaudet AL, Sly WS, Valle D (eds), McGraw-Hill, pp 1195–1214

43. Menzel H-J, Kladetzky R-G, Assmann G (1983) Apolipoprotein E polymorphism and coronary artery disease. Arteriosclerosis 3: 310–315
44. Mitchell JR (1990) What do we gain by modifiying risk factors for coronary disease? Schweiz Med Wochenschr 120: 359–364
45. Mossey JM (1981) Psychosocial consequences of labelling in hypertension. Clin Invest Med 201–207
46. Muldoon MF, Manuck SB, Matthews KA (1990) Lowering cholesterol concentrations and mortality: a quantitative review of primary prevention trials. Br Med J 301: 309–314
47. Newman TB, Browner WS, Hulley SB (1990) The case against childhood screening. JAMA 264: 3039–3043
48. Oliver MF (1992) Doubts about preventing coronary heart disease. Br Med J 304: 393–394
49. Oliver MF, Heady JA, Morris JN, Cooper J (1978) A cooperative trial in the primary prevention of ischemic heart disease using clofibrate. Br Heart J 40: 1069–1076
50. Parthasarathy S, Steinberg D, Witztum JL (1992) The role of oxidized low-density lipoproteins in the pathogenesis of atherosclerosis. Ann Rev Med 43: 219–225
51. Rajavahisth TB, Andalibi A, Territo MC, Berliner JA, Navab M, Fogelman AM, Lusis AJ (1990) Induction of endothelial cell expression of granulocyte and macrophage colony-stimulating factors by modified low-density lipoproteins. Nature 344: 254–257
52. Ramsay LE, Yeo WW, Jackson PR (1991) Dietary reduction of serum cholesterol concentration: time to think again. Br Med J 303: 953–957
53. Rauh G, Keller C, Schuster H, Wolfram G, Zöllner N (1992) Familial defective apolipoprotein B-100: a common cause of primary hypercholesterolemia. Clin Investig 70: 77–84
54. Reckless JPD (1987) Can nutrition favourably affect serum lipids? Proc Nutr Soc 46: 361–366
55. Report of the National Cholesterol Education Program Expert Panel on the detection, evaluation, and treatment of high blood cholesterol in adults (1988). Arch Intern Med 148: 36–69
56. Ross R (1992) The pathogenesis of atherosclerosis: In: Heart Disease. A textbook of cardiovascular medicine. Braunwald E (ed) WB Saunders, pp 1106–1124
57. Schwartz CJ, Valente AJ, Sprague EA, Kelley JL, Cayatte AJ, Rozek MM (1992) Pathogenesis of the atherosclerotic lesion. Implications for diabetes mellitus. Diab Care 15: 1156–1167
58. Scientific Steering Committee on behalf of the Simon Broome Register Group (1991) Risk of fatal coronary heart disease in familial hypercholesterolaemia. Br Med J 303: 893–896
59. Soria LF, Ludwig EH, Clarke HRG, Vega GL, Grundy SM, McCarthy BJ (1989) Association between a specific apolipoprotein B mutation and familial defective apolipoprotein B-100. Proc Natl Acad Sci USA 86: 587–591
60. Stamler J, Wentworth D, Neaton JD (1986) Is relationship between serum cholesterol and risk of premature death from coronary heart disease continous and graded? Findings in 356222 primary screenees of the multiple risk factor intervention trial (MRFIT). JAMA 256: 2823–2828
61. Stason WB (1990) Costs and benefits of risk factor reduction for coronary heart disease: insights from screening and treatment of cholesterol. Am Heart J 119: 718–724
62. Steinberg D, Parthasarathy S, Carew T, Khoo JC, Witztum JL (1989) Beyond cholesterol. Modifications of low density lipoprotein that increase its atherogenicity. N Engl J Med 320: 915–924
63. Study Group, European Atherosclerosis Society (1987) Strategies for the prevention of coronary heart disease: a policy statement of the European Atherosclerosis Society. Eur Heart J 8: 77–88
64. Study Group, European Atherosclerosis Society (1988) Recognition and treatment of hyperlipidemia in adults: a policy statement of the European Atherosclerosis Society. Eur Heart J 9: 571–600
65. Summary of the presentation to the Canadian Consensus Conference on Cholesterol, March 9-10, 1988, Ottawa, Canada. (1988) Paper 10: Logan A. The Conclusions of the Canadian Task Force on the Periodic Health Examination: hypercholesterolemia

66. Tsevat J, Weinstein MC, Williams LW, Tosteson AN, Goldman L (1991) Expected gains in life expectancy from various coronary heart disease risk factor modifications. Circulation 83: 1194–1201
67. Turpeinen O (1979) Effect of cholesterol-lowering diet on mortality from coronary heart disease and other causes. Circulation 59: 1–7
68. Utermann G (1989) The mysteries of lipoprotein (a). Science 246: 904–910
69. Williams RR, Hunt SC, Wu LL, Hopkins PN, Hasstedt SJ, Schumacher MC, Stults BM, Kuida H (1990) Concordant dyslipidemia, hypertension and early coronary disease in Utah families. Klin Wochenschr 68 suppl XX: 53–59

Author's address:
Prof. Dr. G. Schmitz
Institut für Klinische Chemie und Laboratoriumsmedizin
Universität Regensburg
Franz-Josef Strauß-Allee 11
D-93042 Regensburg
FRG

Subject Index

MIX
Papier aus verantwortungsvollen Quellen
Paper from responsible sources
FSC® C105338

If you have any concerns about our products,
you can contact us on
ProductSafety@springernature.com

In case Publisher is established outside the EU,
the EU authorized representative is:
**Springer Nature Customer Service Center GmbH
Europaplatz 3, 69115 Heidelberg, Germany**

Printed by Libri Plureos GmbH
in Hamburg, Germany